In praise of Beijing to Barbados in a Rowboat

"After rowing 5,000km across the roiling Atlantic Ocean with buttocks full of boils and a former PLA soldier who had never been to sea, Christian Havrehed has a kind of story that drops jaws. The beauty of his book is that he knows the story will tell itself.

After living and working in China since 1989, Havrehed wanted a Chinese rowing partner. In Sun Haibin he found a fit long-distance runner who'd spent much of his adult life in Xinjiang – about as far from the ocean as it's possible to get. Havrehed himself had little rowing experience.

Havrehed says one of the keys to their success was remaining friends. Even with his experience in China and his fluency in Putonghua, he became exasperated by Sun's Chinese habit of demonstrating affection by fussing over whether his partner was eating well, wearing enough or using sunscreen. Sun snapped over Havrehed's constant understatement – the Danish fear that being too positive will bring bad luck.

Before long the book will become a manual for adventurers."

South China Morning Post

"Can two near-strangers from vastly different cultural backgrounds, neither of whom have ever rowed before, summon the common understanding needed to survive two months of isolation at sea? Relationships, even marriages, have come spectacularly unstuck during the race.

This interesting cross-cultural aspect is at the heart of *Beijing to Barbados* ... through it all, regular doses of gallows humour, mutual encouragement, friendly competition and the odd burst of euphoria keep the pair paddling steadily westwards, and ready the crew of the *Yantu* for a final, unexpected challenge as they approach the coast of Barbados.

Beijing to Barbados in a Rowboat is an adventure story with a contemporary Asian flavor ... Experienced and armchair adventurers alike will enjoy accompanying the pair through travail to triumph."

The Asia Review of Books

BEIJING TO BARBADOS IN A ROWBOAT
2020 Updated Edition
ISBN: 978-988-97427-0-6

First published in Paperback in 2003
© 2003 Christian Havrehed

Published by Impact Books, a division of Impact Consulting Ltd
Unit 2302, 23rd Floor, 9 Chong Yip Street, Kwun Tong, Kowloon, Hong Kong

Typeset in Cambria by Hammad Khalid
Cover designed by Maja K. (@_m_design3)
Cover photographs by Véronique Havrehed
Interior photographs mainly by Véronique and Christian Havrehed
Illustrations by Lin Qing

Fifty percent of net proceeds from this updated edition will be split equally between all the United World Colleges across the world, for them to use towards scholarships as they see fit.
(The first edition pledged twenty-five percent of net proceeds and resulted in US$1,500 for the United World College of the Atlantic).

UWC

If you cannot find this book in your local bookstore or on-line it can be purchased through *www.yantu.com*

All rights reserved. No part of this book may be reproduced in any form or by any electronic means, including information storage and retrieval systems, without permission in writing from the publisher, except by a reviewer who may quote brief passages in a review. The right of Christian Havrehed to be identified as the Author of the Work has been asserted as have his moral rights with respect to the Work.

BEIJING TO BARBADOS IN A ROWBOAT

CHRISTIAN HAVREHED
黄思远

THE YANTU PROJECT

Impact Books

In memory of

Hui Wang
王辉
* 26.11.1983 † 7.10.2006

Risk

To laugh is to risk appearing a fool.
To weep is to risk appearing sentimental.
To reach out to another is to risk involvement.
To expose your feelings is to risk rejection.
To place your dreams before the crowd is to risk ridicule.
To love is to risk not being loved in return.
To go forward in the face of overwhelming odds is to risk failure.

But risks must be taken for the greatest risk is to risk nothing.
The person who risks nothing, has nothing and is nothing.
You may avoid suffering and sorrow, but you cannot learn, you cannot feel, you cannot change, you cannot grow and you cannot love.
Chained by your certitudes, you are a slave.
Only the person who risks is truly free!

Anon

Contents

Forewords 7
Acknowledgements 11
Preface 13
   You seem crazy enough 20
   Taking the plunge 44
   Things start to take shape 65
   Completing the hull 71
   The China launch 82
   Training and fund raising 93
   Leaving Hong Kong 111
   Tenerife — meeting the competition 128
   The race starts! 151
   A stormy beginning 158
   Through the barrier 174
   Heading west 184
   The last thousand miles 202
   Sprint to the finish 224
   Hanging out in Barbados 233
   Back to China 240
   Postscript 243
   Appendix 248
Afterword – 19 years on 253
Support Christian's adventures 263

Forewords

By the President of United World Colleges

This is a story of a great adventurous challenge. It is a story of survival and success against the odds. It is a story to excite us. As President of the United World Colleges, it is my great pleasure to introduce the story because Christian Havrehed is a graduate of the United World College of the Atlantic in Wales.

The United World Colleges is a global educational movement, which brings together students from diverse socio-economic and cultural backgrounds in an environment designed to foster international understanding, peace and tolerance. In the 'Yantu Project', Christian was able to combine these values, his love of the Ocean and of adventure and his interest in China. He also decided to use the project to raise the profile of the UWC movement and to raise funds to enable Chinese students to study at Atlantic College.

The Yantu Project was an audacious undertaking in which Christian entered the Atlantic Rowing Challenge with his rowing partner Sun Haibin in order to become the first Dane and the first Mainland Chinese to row across the Atlantic. Christian's story is a compelling account of the challenges faced by two strangers from very different parts of the world. It charts the development of a dream, the challenges of having a boat built in China, the detailed preparation needed, the demanding training schedule and the many obstacles they had to overcome.

Her Majesty Queen Noor of Jordan

The book paints a graphic picture of a journey of 56 days and 2,745 nautical miles during which Christian and Sun Haibin suffered severe discomfort, solitude and personal risk. At the same time it illustrates the determination needed to make the dream become reality and gives an insight into the unique friendship and understanding which developed across cultural boundaries.

The promotion of international understanding is the very foundation on which the United World Colleges movement is built and this book sends us a powerful message — that it is possible to achieve the seemingly impossible so long as there is trust in one another and a shared vision. In a world where we are surrounded by tension and misunderstanding, this message should encourage us all.

By the Secretary General of the Chinese Yachting Association[1]

Quanhai Li (李全海)

In December 2000, an invitation hailing from the Atlantic coast caught our attention. It came from a Dane who hoped that the China Water Sports Administration could help find him a Chinese rower with whom he could row cross the Atlantic Ocean in a double-handed rowing race. The Dane wanted to use the race to raise scholarships for Chinese students to study overseas at the United World College in Wales.

The English name of this Dane was Christian Havrehed. He had a splendid sounding Chinese name, 黄思远, "Huang Siyuan" – "Far-thinking Havrehed". He presented us with a plan that was full of enthusiasm and idealism: "Firstly, it will provide the opportunity for the first Chinese national to row across an ocean. Secondly, China taking part in an international competition like this will attract a lot of international media attention ..." He boldly added, "It will be a great project completely based on goodwill. I cannot offer any overseas trips or banquets, but I undertake to meet all race costs and guarantee the proper undertaking of all matters concerning the race. I will also cover the rower's expenses in connection with travelling to Hong Kong for training..."

His project proposal could move the heart of almost anyone with dreams, including ours. Before this, the China Water Sports Administration had never tried to organize a project with an individual, but this project was based on the love of Sino-foreign cooperation and oceanic exploration and was without a doubt worthy of support. Thus we decided to help him as far as possible with realizing the project, though at the time it seemed more like a dream.

We found Zhang Jian who at the time had already succeeded in swimming across the Bohai Strait and was China's most famous long-distance swimming champion. Zhang Jian was an outstanding

---

[1] Foreword translated from the Chinese version of the book "划越: 一个中国人和一个丹麦人横渡大西洋的故事", published in 2004.

athlete and also Secretary General of the Beijing Ironman Association. He recommended Sun Haibin to us, a student from Beijing Sports University. Sun Haibin had placed third in the individual event at the Asian Ironman Competition. His physical condition and mental state were very suitable for the adventurous activity of crossing the Atlantic Ocean in a rowboat. Sun Haibin and Christian were very well matched and they decided to undertake the project together.

In a way, Christian and Sun Haibin's story is legendary. Two individuals from different cultural backgrounds spent over 50 days together in the middle of the ocean, day and night, in a space less than three square meters. Furthermore, before undertaking this project the two of them did not know each other and Sun Haibin had virtually no rowing experience. Would they be able to stand the loneliness? Would storms and whales sink or swallow them? All of this was unknown. But as a result of Christian and Sun Haibin's painstaking preparations and efforts for 56 days out at sea - and to the surprise of many people - they succeeded in rowing across the Atlantic Ocean.

This was a case of perfect cooperation between the culturally different China and the West. The scholarship money raised from the project was all used to support Chinese students to study overseas. It was a highpoint of human endeavor that embodied mankind ideals of pursuing unity, peace, and common development. At the same time, for China's non-government maritime activities, it was a successful experience of opening up our door and stepping into the world.

As a result of having participated in this project, we will now strongly support this type of international maritime cooperation going forward.

We are convinced that readers will cherish this book. That their legendary story has been successively published in English, Chinese and Danish already proves that the ideals of "cooperation, adventure and compassion" engage people across nationalities.

This book will tell you a true story of beauty and hardship. I wish it every success.

Acknowledgements

This book has been a long time in the making. Long, because before being able to write this book, I had to build a boat in China, find a Chinese rowing partner and row across the Atlantic. It would therefore be inappropriate to only acknowledge people who have been involved in putting this book together. Without the boat there would have been no book and the following 'thank you' is therefore dedicated to everyone who got involved and helped out along the way.

I would like to thank the China Rowing Association, The Hong Kong, China Rowing Association, the Royal Hong Kong Yacht Club, Keith Mowser, Luyang Boat Building Company, and Zhang Jian for helping getting the boat built and introducing my rowing partner Sun Haibin. You probably all thought I was mad, but without your assistance there would have been no boat, no crossing and no book.

John Lawrenson and Malcolm McKenzie from the United World College of the Atlantic kindly provided the endorsement legitimising my fundraising campaign to send Mainland Chinese students to study at Atlantic College, and ex-UWC students Francine Kwong, Tammy Wan and Michael Yong-Haron helped raise scholarships. Robert Bird offered the services of the Royal Hong Kong Yacht Club as custodian for funds raised and Peter Davies wrote endless letters to potential sponsors. Inge Strompf-Jepsen provided many valuable leads and Kate Vernon set up lunch talks with several Rotary clubs, securing the support of Rotary Club Hong Kong North East and Rotary Club of Kowloon North.

A big thanks to DelfiTech for becoming the first cash sponsor, to Hans-Henrik Madsen and Pernille Kragh Rühe of Viking Life Saving Equipment, and to Stratos for providing Iridium communications equipment and air time. Also thank you to Jebsen, Jardines, Sallmans, Vickers, Bruce Bowers, Bill Benter, Olav Storm and numerous other friends, family and supporters who contributed towards scholarships and race costs.

Without expert assistance from Rob Hamill, Bob Wilson, Martin Reynolds, Rob Stoneley and Hong Kong Olympic rowing coach Chris Perry I doubt we would have known what direction to face in the boat, how to rig the rowing stations, and how to row — at least to some degree of competence! Dr Brian Walker provided medical advice which helped us protect and repair our bodies.

The late Judy Nip, Duthie Lidgard, Mr Yip, Ah Bun, Moon, Connie Chan and the rest of the Royal Hong Kong Yacht Club Boat Yard gang took great care of *Yantu* and equipped her to the highest standard.

On the PR side big thanks are due to Wang Zengshuan and Yu Zhen from CCTV in Beijing and Victoria Button from the *South China Morning Post* in Hong Kong who were the first to pick up on the story and reported on it faithfully to the end. Inge Nielsen provided a place to live when I was in Beijing and introduced me to Sun Hongmei, who did the most fabulous translations of our website until she went to the US to study. Rickie Tsui helped coordinate press conferences and Lewis Lui and Perry Tam helped build the Yantu.com website. Many thanks to my editor Alan Sargent, who provided patient and constructive challenge to the manuscript, 'Big Al' for the photo layout and Keith Clark from the United World College International Office for arranging the foreword. For this second edition additional thanks to Hammad Khalid, Maya K, and Clemens Schulenburg.

Throughout the project I was ready to throw in the towel and burn my boat at least once every month because things were just getting too hard and, had it not been for my wife Véronique, I might well have done so. Thank you for your unwavering support!

Finally, thank you to all the well-wishers, supporters, and sponsors, who sent messages of encouragement, advice and congratulation, of whom there are too many to mention all by name. Your messages kept us going when things were tough and helped make our scholarship fundraising a success.

Preface

It is now 19 years ago Sun Haibin became the first Chinese (and Asian) to row across an ocean and I became the first Dane. As a result, Sun Haibin also became the first amateur athlete to be nominated for Sportsman of the Year in China. Our project was trail-blazing and it went (pre-social media) viral. We became the first privately organized amateur sport project to be supported by the Chinese government and featured on National Chinese Television. It was a project way ahead of its time. Perseverance, visioning, can-do attitude, humour, and cross-cultural cooperation made it possible.

In 2001 we competed as a pair in the 2nd Trans-Atlantic Rowing Race in history. When we set out from Tenerife less than 100 people had successfully rowed across an ocean and 7 had died trying. Now more than 1,100 people have successfully rowed across an ocean, and only 3 more have died trying, so safety statistics have improved dramatically. Despite the 10-fold increase in ocean crossings, ocean rowing remains a niche sport. More people have climbed Mt. Everest than rowed across an ocean.

The reason for republishing this book is to provide the reader with some perspective on how much China has changed in just 19 years. This is mostly done in this Preface and the Afterword. The rest of the book is essentially the same as the first edition, but with a different layout, some new inserts, and better picture quality.

The pictures in the first edition were grainy because back in 2003 it was prohibitable expensive, if not impossible, for the designer in London to email large picture files to the printer in Hong Kong using dial up internet, so we had to settle for less than perfect. The pictures in this edition are better, but still not perfect, as common cameras then also were not as good as now, and I have not been able to locate all the 35mm negatives. Negatives! What's that? Well, exactly. How times have changed. Not just with respect to speed and cost of information sharing, but also in terms of how information gets disseminated. The advent of the internet with social media platforms, like Facebook and WeChat (China's answer to Facebook), means TV and newspapers have become less influential.

Chinese society has changed too, and so has the way the world engages with China.

Back in 2001 it was difficult for the Chinese to go abroad on holidays. Firstly, they did not have much holiday. Secondly, they hardly had any money, and thirdly, because they had little money many countries were reluctant to grant Chinese visas out of fear that they might never leave again. That has all changed. Now the Chinese have both spare time and money so countries are now going out of their way to make it easy for them to visit – and spend money. If someone had said back in 2001 that by 2020 Charles de Gaulle airport in Paris would have signage in Chinese, people would have laughed at the idea, but that is in fact what has happened. And not just in Paris. The luxury stores in most major European cities now employ Chinese-speaking staff to better communicate with their highest spending customers. The trouble we had getting Sun Haibin out of China in 2001 is a non-issue today.

When I first went to China in 1989 as a student backpacker, I felt like one of the richest people there, though I had hardly any money. When I left in 2013, I was a well-paid Managing Director of an international company, with company car, driver, maid, and a 358m^2 villa within a gated compound. However, I was living in the poor end of the compound where the expats huddled together. The upmarket end of the compound was where the Chinese lived in their 1,000+m^2 villas, complete with internal elevators and large garages to park their Ferraris and Maseratis. So, although I was much better off in 2013 than in 1989, relatively speaking I was much poorer. In 2001 I paid for all our race expenses, because it was inconceivable that Sun Haibin or any other average Chinese person would have money to spend on an adventure like ours, but now plenty of Chinese have plenty of money. With our project we raised money to send Chinese students overseas to study at a United World College because they could not afford to pay themselves.

Now China has more students studying overseas than any other nation and the Chinese are paying their own tuition fees. There is now even a United World College in China and foreign students are increasingly going to China for international education and work exposure. Young people now strive to have work experience in China on their CV.

The fundraising focus with this second edition has therefore changed. In the first edition I donated 25 percent of net proceeds to send more Mainland Chinese students to Atlantic College, but since China is now well represented at all United World Colleges around the world, 50 percent of net proceeds from sales of this second edition will be shared equally among the United World Colleges, to help them finance scholarships, as they see fit.

Back in 2001 when we fundraised for scholarships, we did not get one single penny from individuals or corporates in China because they had only just started making money and were therefore spending it on themselves. Moreover, fundraising logistics and ethics were cumbersome to manage because on-line fundraising platforms did not yet exist.

However, now Mainland China has more US dollar billionaires than any other country and the superrich are starting to see philanthropy as a way of building a positive image and legacy for themselves, just like Rockefeller and Gates. On-line fundraising platforms are now commonplace.

Who would have thought that since China entered the WTO in 2001, Mainland Chinese entrepreneurs, starting from practically zero, would have been able to amass such wealth that they now feel in a position to donate USD millions and even USD billions to charity? If someone in 2001 had said that in 2020 a Chinese billionaire would donate medical supplies to help Western governments deal with a global health pandemic, people would have laughed. But that is exactly what Alibaba's Jack Ma has done.

In 2001 the Chinese seemed unsure about their own worth and Westerners who went to work in China were automatically granted a 'Foreign Expert' visa simply because they were foreign. As long as you were white, and non-Russian, you would be treated with respect and listened to, regardless of whether you deserved it. That has changed.

When the USSR collapsed in 1989, China won a 33-year old ideological battle with the Soviet Union about who had the better form of Socialism. Socialism with Chinese characteristics was further vindicated in 2008 when the Western world screwed up the Global Economy royally with its unchecked Capitalism and caused a major global recession (that China weathered better than most). As a result, democracy and capitalism have lost some of their shine and China is now less likely to listen to the West.

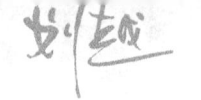

The Chinese economy has also helped boost China's national pride. In 2001 China had an economy about the size of France's. Now it is the world's second largest economy, almost three times larger than Japan's (the previous number 2), and 2/3 the size of the USA's, the world largest economy. In just 19 years China has become the world's second economic superpower, inadvertently challenging the USA's domination. China has also started to invest heavily in Africa, a resource-rich continent that the West has shown limited strategic interest in developing (might change now that China is there), and China is also pushing through with its Belt and Road Initiative, which will secure logistics and political influence through Central Asia and the Middle East all the way to Europe.

The pride in being Chinese and the belief that the Chinese can hold their own against anyone has also been boosted by such events as the Beijing 2008 Olympics and 2010 Shanghai World Expo, as well as individuals like NBA basketball player Yao Min and hurdler Liu Xiang.

When Sun Haibin and I participated in the 2nd Trans-Atlantic Rowing Race in 2001, China and the Chinese were at a completely different stage of development compared to now. Sun Haibin told me we were way ahead of our time and that at least 15 years would pass before someone else would try to do something like this from China. He turned out to be right. It was only in 2017 that China again entered a boat into the 12th Trans-Atlantic Rowing Race. *Kong Fu Cha Cha* was rowed by four Chinese girls from Shantou University. Arriving in just 34 days they set a new world record for the fastest female 4 crew entry, bettering the 2016 British record by 6 days!

Shantou University is sponsored by the richest person in Hong Kong, Li Ka Shing, so the girls' row was not paid for by Mainland Chinese sponsors, but there is no doubt that there is more and more appetite amongst Mainland Chinese to sponsor such initiatives. A litmus test could be my next China-related nautical adventure introduced in the *Afterword – 19 years on*.

The girls raised RMB16,249.53 (USD2,300) for a charity called "Teach for China" from Mainland Chinese donors. That is not a lot, but it is a start. I like to think it is Sun Haibin and me who inspired them to attach an education charity to their rowing campaign.

Rowing across the Atlantic, Sun Haibin and I had the mantra "Anything is possible". China, with its astronomical socio-economic development, its superior 5G technology, and the world's first (and still only) landing on the dark side of the moon, proves just that.

That is good news for China and painful for the Western world. The world order the West takes for granted has been upset and a new equilibrium is slowly emerging with China as a key player. Whether we like it or not, the Western world will have to come to terms with this. That will not be easy.

Sun Haibin and I had written in big letters over our cabin hatch "合作可以更多", which translates into "Together we achieve more". That mantra, and respect for each other's differences, made us ultimately succeed as a team and become friends for life. Hopefully, the world's leaders will also choose to get into the same boat and embark on a similar journey of trust and mutual understanding. I find it slightly difficult to picture the presidents of China and the USA rowing butt-naked together in the middle of the Atlantic Ocean, but then again ... *Anything is possible!*

The text between Sun Haibin and I translates "Together we achieve more". Had we forgotten this we would never have succeeded.

Beijing to Barbados in a Rowboat

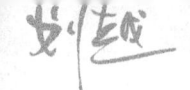

Route of *Yantu*

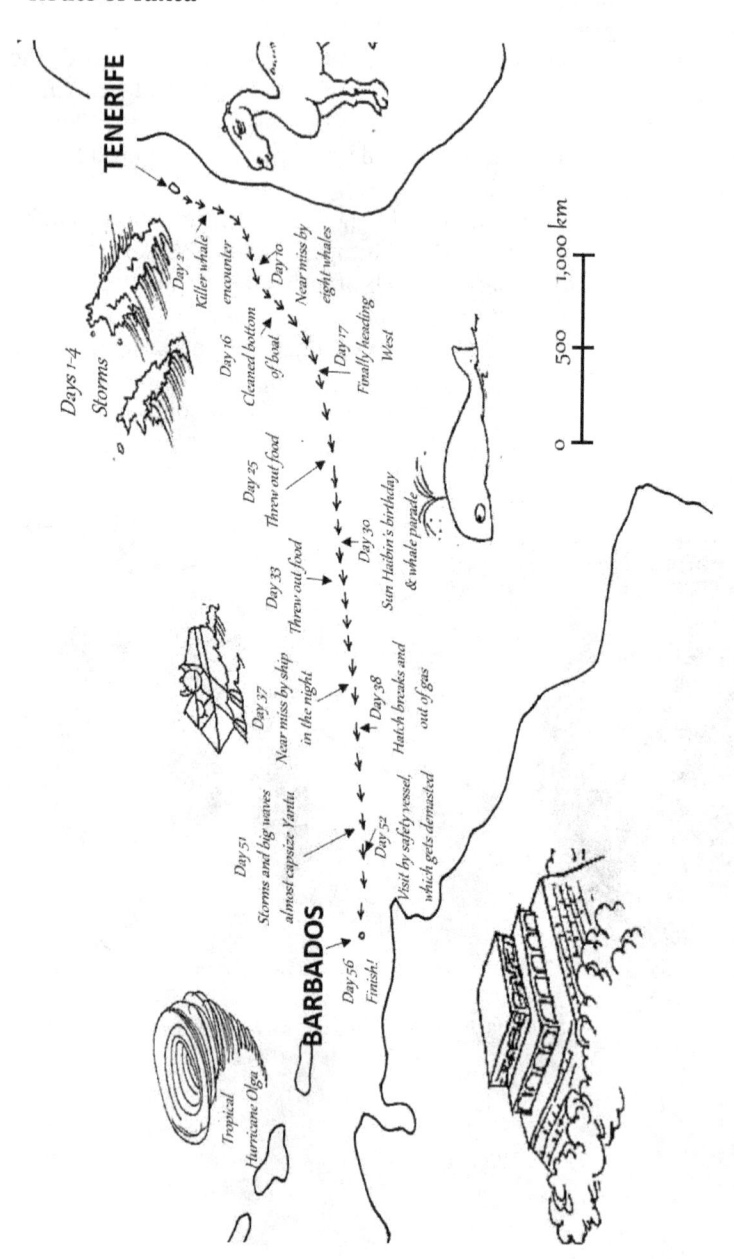

The anatomy of the Ocean rowboat *Yantu*

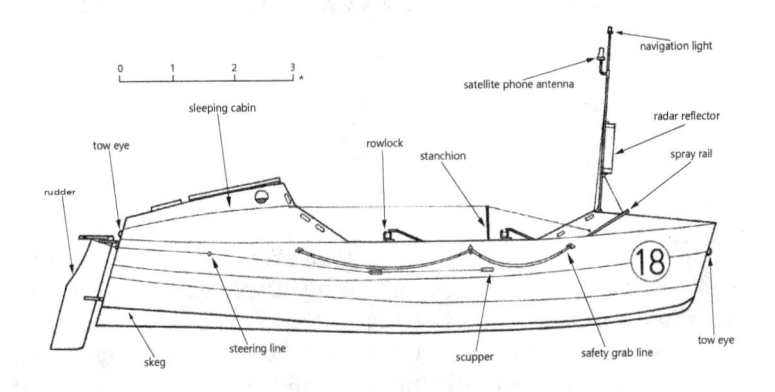

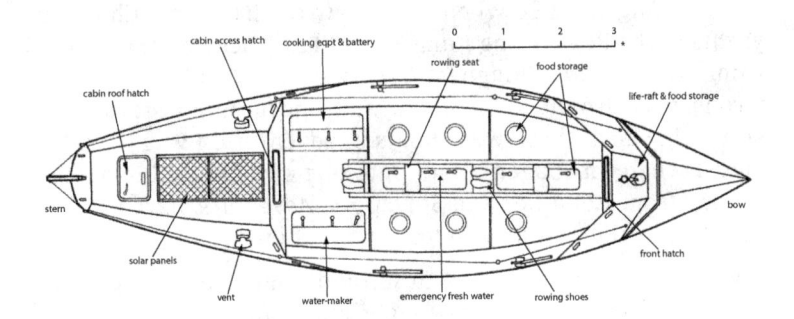

You seem crazy enough

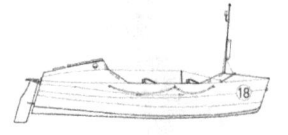

'Why don't you do this instead? You seem crazy enough to do it,' said skipper Andy Dare, extending a brochure in my direction. I was lying on the deck feeling miserable, not from seasickness, but rather from boredom and too much sun. The engine was puttering away and that was the heart of the problem.

It was August 1998 when I first saw the BT Global Challenge yachts at the Royal Hong Kong Yacht Club. These were the boats from 'The World's Toughest Yachting Race', which sailed the wrong way around Cape Horn and encountered storm after storm. In short, these were boats of adventure! I was therefore elated when I managed to get onto one of them sailing a delivery leg from Hong Kong to Singapore. *Adventure, here I come*, I thought, applied for annual leave and off I went.

We were now six days out to sea and somewhere off the coast of Malaysia making a steady eight knots under motor. We had been motoring since leaving Hong Kong because of lack of wind. I do not think much of motor-sailing, that is for people who are motivated by showing off their kit and not much else. Real sailing is different, you do it to be immersed in the elements and to get a good work-out. So lying on the deck of a sailing boat under motor, sunbathing for the sixth day in a row with nothing to do was a big disappointment. Feeling lethargic, I took the brochure from Andy.

WARD EVANS CROSS ATLANTIC ROWING RACE — THE WORLD'S TOUGHEST ROWING RACE, was printed across the front page along with a picture of a white outsized rowing boat crewed by two fit-looking rowers. Andy continued: 'If you do that you will never get bored. You will have to row all the time. Ha, ha!' With that remark he went back below deck. I sat up and looked closer at the brochure.

I had only ever rowed twice in my life. The first time I was about eight years old and my family had been visiting friends living on a fjord in Denmark. When you are eight you get bored quickly, particularly when your parents are talking to friends about nothing you can relate to, so eventually, to get some peace and quiet, they sent me and our friends' son down to the beach and told us we could go for a row in their tender as long as we stayed close to shore. It turned out that the adult's and child's definition of 'close to shore' varied significantly and my parents were not impressed after we returned from the shipping channel a few hours later!

Inspecting blisters after returning from my first ever row. I luckily forgot that painful experience, or I would never have set out to row across the Atlantic.

The second time I tried rowing was during freshman week at Durham University in 1989. I thought it would be a good way to keep fit. Most of the students trying out had posh English accents, seemed to long for the return of the Empire and appeared motivated to row because that was what one did at Oxford and Cambridge where they wished they were studying. With my lack of Imperial pedigree and Danish accent I did not fit in very well so I never went back.

Yet the brochure intrigued me. What struck me the most was the simplicity of the whole thing. Build a small ocean-going rowboat, find a rowing partner and then row unsupported 3,000 miles from Tenerife to Barbados. *How difficult can that be?* I thought, and there and then I decided I was going to do it.

'Andy, how do I sign up?' I yelled down the hatch. 'Have you had too much sun?' the reply came back. 'No, no. I am dead serious. I've got to do this.' Andy came back up on deck and told me about the race.

Like the BT Global Challenge it was organised by the Challenge Business, a company set up by Sir Chay Blyth. Chay wrote himself into yachting history by becoming the first person to single-handedly circumnavigate the world against the prevailing winds and currents non-stop. I had read Chay's book *British Steel* about his 'wrong way round' and was aware that he had subsequently modelled the BT Global Challenge on this achievement.

However, I was not aware that he had also rowed across the Atlantic in 1966 together with John Ridgway, his then superior in the army. In fact, it had never crossed my mind that anyone could row across an ocean.

But back in 1966 Chay and John had got into their rowing boat *English Rose III* in Orleans, Cape Cod, USA, and rowed across to the Aran Islands, Ireland. They had rushed their preparation because they did not want another UK challenger, David Johnston and John Hoare, rowing a boat called *Puffin*, to become the first to row across the Atlantic since Norwegians George Harbo and Gabriel Samuelsen had rowed across in 55 days in 1896, the first people ever to do so — at least in modern history I hasten to add, out of fear of offending the Vikings and Irish monks.

You seem crazy enough

In 1896 Norwegians George Harbo and Gabriel Samuelsen rowed across the Atlantic in 'Fox' in 55 days. Safety equipment amounted to life belts filled with reindeer hair.

Rumour has it Chay could not row at the outset and journalists were shaking their heads in disbelief at his lack of skill when they asked him for a rowing demonstration prior to setting off. He certainly learned it along the way and they completed the row in 91 days. However, Chay and John's friendship did not last the pressure-cooker experience of the crossing. David Johnston and John Hoare in *Puffin* were not as lucky and were lost at sea. *Puffin* was later recovered.

Chay Bligh (left) and John Ridgway leaving America in 'English Rose III' in 1966.

Historically, rowing across the Atlantic has not been a popular pastime, and until the first Atlantic Rowing Race in 1997, only 39 attempts had been made. The safety statistics were not good either, with five rowers lost at sea. There was therefore a lot of scepticism when Chay decided to launch a Cross Atlantic Rowing Race in 1997, again modelled on his own achievement. However, the race was a huge success. Twenty-four boats out of 30 entries completed the crossing, in spite of plenty of close shaves no one was lost at sea, and Rob Hamill and Phil Stubbs from New Zealand won the race in a staggering 41 days, thereby finally breaking Harbo and Samuelsen's record set 101 years previously!

The brochure in my hand was advertising the 2001 Ward Evans Cross Atlantic Rowing Race. It was 1998 and I therefore had plenty of time to prepare. The 1997 race had significantly improved safety statistics. The lost-at-sea ratio had dropped from five in 39 attempts to five in 69 and considering that boat design and safety is ever improving I thought the odds were not too bad. And it would be a real adventure that I could control, instead of simply just participate in.

This was important to me because, on our way to Singapore, we were discussing the pros and cons of participating in the BT Global Challenge as a paying crewmember. It became apparent to me that the best performing boats in the 1996–97 race had assigned crewmembers to the tasks they were best at, as opposed to letting everyone have a go at all the different roles on board. So some people had paid around US$40,000 to basically cook their way around the world.

One of my six sunbathing crewmembers was driving me crazy. Sharing company with her until Singapore was definitely long enough and if I paid to participate in a round-the-world ocean race I could again end up with a fellow crew member I did not get on with. One person would be OK, but how about five or six? In joining a big crew of 17 there would be no guarantees. However, if I chose to row across the Atlantic I would be able to select my own partner, thereby greatly improving the odds of having enjoyable company. As I was sitting there baking in the sun listening to Andy talk it never occurred to me that, just maybe, it might not appeal to that many people to row across the Atlantic!

Finally, it would be a great challenge to manage the race entry and shape it in any way I wanted to. It would be a true and fair challenge — I would either be able to make it or not, and it would be entirely up to my own efforts. It would not simply be a matter of paying money, showing up and being more or less guaranteed success, as the feat of climbing Mount Everest has increasingly turned into.

For all those reasons I thought rowing the Atlantic was a wonderful idea and when Andy told me there were still free spaces I was hooked. We arrived in Singapore on a Wednesday and I immediately called Teresa Evans at the Challenge Business in Plymouth and paid my deposit by credit card. All done, I said goodbye to my fellow crewmembers and headed for Koh Samet in Thailand to learn how to scuba dive before heading back to Hong Kong for work the following Monday.

In between diving trips a plan for rowing the Atlantic slowly began forming in my head.

* * * *

I grew up in Denmark, spending all my holidays on a small island called Læsø. This is a great place for a child. The island only had one policeman who had to live in a close-knit society of mainly fishermen and farmers and therefore could not be too strict. Læsø was the perfect place for a 10-year-old to go hunting with a shotgun and drive mopeds, way below the legal age limit.

I purchased my first dinghy when I was seven for the astronomical sum of US$20. No one in my family knew how to sail, so I learned by trial and error. The dinghy was small, homemade, and not particularly seaworthy. It sank when it got full of water. Going through the surf I sometimes got a wave on board and found myself at the bottom of the sea. I then had to ask the occasional grownup swimmer to lift the boat up to the surface with me in it so that I could bail out the water and continue my adventures.

I purchased my first dinghy 'IMP' for US$20. She was not particularly seaworthy and prone to sinking, but on the upside she was so small I could easily tow her on a trailer behind my bicycle.

Through sailing on Læsø I made friends with Olav Storm and we became great sailing buddies. We pretty much went out in all kinds of weather and that was how our fathers met. One worried father standing down at the harbour looking out to sea through a pair of binoculars trying to spot his son's dinghy in the thunderstorm and approaching fog is approached by an equally worried-looking father with binoculars. Father # 1 to father # 2: 'My son is out there.' Reply: 'So is mine.' 'Oh, you must be Olav's father then.' The state of the weather seemed to exclude the possibility that it could be anyone else but Olav and me out there. We eventually made it back safely. Later on Olav and I lived on a 28-foot yacht with no standing room for two months in the middle of winter with snow on deck while he was looking for a room to rent and I was saving up money to travel around Asia. It was horribly cold and cramped, but we thoroughly enjoyed it.

During my holidays I also spent a lot of time preparing firewood for the winter. Our house bordered on a small forest and every summer I would prepare about 20 cubic metres of firewood for the winter. This involved cutting down trees with a chainsaw, then de-branching them, cutting the trunk into suitable sized logs, loading them on a wheelbarrow, pushing the wheelbarrow back through the forest and then stacking the logs neatly close to the house for easy use in the winter. This was an extremely good workout from which I gained a lot of strength and endurance in my upper body. To speed up the logging, I would often work with friends and we would cut down the trees simultaneously. One unlucky day, as I was preoccupied with watching my own tree fall, my friend's tree fell right onto me and knocked me down. Apparently, my father arrived on the scene 'at the speed of an ambulance on two legs'. Luckily, I was only hit by the branches at the top of the tree and not the stem of the tree itself. The gash in my neck was not very deep and after stopping the bleeding we continued working.

Preparing firewood for the winter – an excellent way of building stamina useful for rowing the Atlantic.

Ever since I learned to read, I read about adventures at sea and the poles. At the age of eight, and deeply impressed by Thor Heyerdahl, a friend and I spent all our pocket money buying provisions for a timber raft trip from the mainland of Denmark to Læsø, a 50-kilometre trip. We never built the raft, but we cooked a lot of Spam on our Trangia stove in the garden fantasising about the upcoming adventure. In winter we would cross-country ski to an old quarry on the outskirts on our village and imagine that we were undertaking a polar trip, just like Amundsen, Scott, Shackleton, and Nansen.

At the age of 16 years I came across a fantastic opportunity, which changed my life. My mother introduced me to an international school called United World College of the Atlantic based in Wales and asked me if I was interested in applying. I looked at the material and was impressed.

Atlantic College was founded in 1962, by the joint efforts of Earl Mountbatten of Burma, Air Marshall Sir Lawrence Darvall, who had been Commandant of the NATO Defence College in Paris and educator Dr Kurt Hahn. The aim of the school is to promote international understanding by bringing together more than 300 students aged 16 to 18 from as many different countries as possible and then let them live and study together for a two-year period. Her Majesty Queen Noor of Jordan is president of the United World Colleges institution and there are now 18 schools around the world. Queen Elizabeth II is the Patron of Atlantic College. Selection is on merit regardless of colour, religion and social status. What counts is to get students who are interested in living in and contributing to an international environment and once identified they are offered scholarships to cover their board and tuition. The teaching philosophy is the brainchild of Dr Hahn, who also founded Outward Bound, Salem School in Germany and Gordonstoun School in Scotland. He believed 'there is more in you than you think' and therefore designed a curriculum which includes academics, activities, and compulsory community service.

> UPDATE
>
> The UWC mission has been updated in recent years to take climate change into account. It now reads: "UWC makes education a force to unite people, nations and cultures for peace and a sustainable future".

Leaving home and meeting students from all over the world certainly had its appeal, but I think it was the community services that tipped the balance. Atlantic College is a certified RNLI (Royal National Lifeboat Institution) station and I would be able to choose my community service from beach rescue, lifeboat rescue or cliff rescue. By looking at the map I could see that Atlantic College was halfway up the Bristol Channel and would be subject to the weather in the North Atlantic. The tidal range was 12 metres. In Denmark there is no tidal range to speak of so North Atlantic waves and 12 metre tides sounded like adventure.

I applied in 1986 and was lucky enough to be awarded one of two scholarships offered in Denmark for Atlantic College that year. It was only when the train pulled away from the station and I waved goodbye to my parents and friends that it struck me I would not be back for a long time. I had to struggle hard to fight back tears, but after a while excitement set in about the unknown lying ahead.

Arriving at Atlantic College was fantastic! The school is in an old castle in green and exposed surroundings leading down to a dramatic seafront with 15-metre high cliffs. The atmosphere was also unbeatable since every single student had tried very hard to get into Atlantic College. No one had been sent there by their parents and would rather be somewhere else. Everyone was incredibly positive and open to everything. I was put in Lawrenson House, which had 24 boys sharing dormitories of four on the ground floor and 24 girls upstairs. There was a common room at the end of the house where we could all hang out.

Aerial view of Atlantic College. Situated in an old castle right on the Bristol Channel.

I met my dorm mates. Robbie was from the Isle of Man, Koh was from Japan and Khramis from Oman on the Arabian peninsula and we were going to live together in a 25-square-metre room. Not your average school experience! Language was an immediate problem, particularly for Koh.

'Koh, how are you?' I asked.

'You?' Koh repeated, looking puzzled and started thumbing his dictionary. His face suddenly lit up.

'Fine,' he said face beaming, having cracked the difficult question.

'How was your flight?' I said, trying to strike up a conversation.

'Flight?' Challenging on both sides, but it taught us patience and six months later the problem had solved itself.

I had a reasonable command of English on arrival, but I still found classes difficult to follow. I would be listening to the teacher and suddenly I would learn a new word. *That is a good word*, I would think and then try and figure out how to spell it, by which time I had forgotten the sentence I was trying to write down. So my notes from the first term are pretty incomplete.

Atlantic College was the first school to pioneer the now internationally recognised International Baccalaureate and one of the first things I had to do after arriving was to pick my subjects. Having finished my selection, I looked down the list and thought: *Hang on, I could have studied all this in Denmark. I should at least study something I couldn't study in Denmark.* So I ran my finger down the list of subjects on offer, looking through A, B, C ... 'Chinese Studies', I read.

I don't know anything about China, so why not study that, I reasoned, updated my subjects, and handed it to the Director of Studies. Fluke decision? Maybe not!

After I moved to China in 1990 my mother told me that my great, great, great, great grandfather Pierre Paul Ferdinand Mourier had been stationed in Macau as the supercargo for the Danish East India Company from 1770 to 1785. Macau belonged to Portugal and served as a hub for foreign traders doing business with China. Foreign traders were only allowed to travel into Guangzhou from Macau for the summer half year where they had to live in a small enclave dedicated to foreigners called the "Thirteen factories". They were not allowed to leave this enclave, so they resided there on top of their warehouses, called "factories". Trade was conducted with China through Chinese middlemen called *hongs*. Foreign traders could not bring their families to Guangzhou, only men were allowed. Wives and children had to stay in Macao. This meant that families were separated for six months each year. As a result of happy reunions after long separations, and also due to the norm of the time, Pierre and his wife Elisabeth had several children during their 15 year stint in Macao. One of these, Adolph Ferdinand Mourier, died at the age of 10 months. If you visit the Old Protestant Cemetery in Macau today you can find his headstone. During his stay in Macao Pierre learned to read and write Chinese. This was an odd thing to do at the time since Chinese and foreigners were purposefully kept apart and it was illegal under Chinese law for foreigners to learn Chinese. Yet Pierre persevered. As a result, Pierre has been recognised as Denmark's first sinologist and his Chinese exercise books can be found at the Royal Library in Copenhagen.

My great, great, great, great grandfather Pierre Paul Ferdinand Mourier who was stationed in Macau from 1770 to 1785.

I also learned from my mother that my grandfather, Captain Christian Fredrik Denys Mourier, sailed with the Danish naval ship *Valkyrien* to the Far East from 1899 to 1900. This was at a time when the saying 'join the navy and see the world' was still true. The navy in those days had an important diplomatic purpose and *Valkyrien* visited Shanghai to promote Sino-Danish trade.

It therefore seems that my fluke decision to select Chinese studies was actually a genetically influenced choice. I just did not know it at the time!

Visiting my relative Adolph Ferdinand Mourier's grave at the Old Protestant Cemetery in Macao. Adolph died 26th August 1776, at the age of 10 months.

Unlike my academic subjects, I was pretty sure about my community service before I arrived. I was certain I would be joining the lifeboats, but in the end I opted for beach rescue.

'We offer beach rescue because it wears you out and thereby keeps your hormones in check,' my Danish teacher, who happened to be Norwegian, told me.

Well, beach rescue was pretty tough. We started out in the pool swimming endless laps. I was always the slowest because I had poor technique, but when we got into the surf and the sea I would pull away because brute force is more important there. We would also kayak in the surf. Our community service comprised two parts: patrolling close by public beaches during the summer and being on standby to conduct rescues when someone had been cut off by the tide, in case neither the lifeboats nor cliff rescue could get to them. The toughest thing we had to do was to practise 'run-swim-run' in winter with snow on the ground. By the time we got out of the outdoor pool everyone who was normally white was blue and everyone previously black had turned white from the cold. I am not kidding! The highlight of my beach rescue career was when our team won both the Welsh and British championships in Surf Life Saving in 1987.

Working hard as a lifeguard. The amount of posing required can be exhausting.

Despite, or maybe thanks to, being self-taught I ended up as the sailing captain at Atlantic College as this called for all-weather boat handling skills. The 12-metre tidal difference and the often huge Atlantic waves bouncing back off the cliff coast would create very confused seas and extreme conditions. On days when it was blowing a force 6-7 only very few of us were allowed out. Waves would then be so big that when you were in the trough of a wave the side of the next approaching wave would completely block the wind, despite the mast on the dinghy being 4.5 metres tall. The landings would then always be exhilarating since you had to time it perfectly in between two breaking waves or risk breaking the boat in the surf.

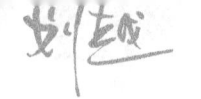

During my two years at Atlantic College I made good friends with a number of Hong Kong Chinese, overseas Chinese and the only Mainland Chinese, Yang Boning. Despite being the most populous country in the world there was only one Mainland Chinese student at Atlantic College because there was only scholarship funding for one student. Typically, each country's government, private industry, and wealthy individuals sponsor a few scholarships, but not in the case of China. Bringing a Mainland Chinese student to Atlantic College therefore depends on the goodwill of other governments and individuals. I always thought it was important to get more Mainland Chinese students to Atlantic College because how can you have truly international understanding if one of the world's most prominent countries is under-represented?

When I graduated from Atlantic College in 1988, I thought I would follow in my grandfather's footsteps and join the navy. However, I was not impressed with the recruiting process and at the time the Danish navy was very small, cutting back, and there was therefore little reason to expect that it would ever be put to any use. Joining the navy therefore sounded somewhat boring. Also, the navy would give me only two weeks of holiday per year for the next eight years, something I was not capable of picturing. I then thought of becoming a boat builder, but in the end opted to study an Asian language. It was a toss-up between Japanese and Chinese. At the time Japan's economy was booming and China was only just starting to open its doors to the outside world. China therefore sounded a lot more like adventure than Japan, and I therefore decided to study Chinese. Since I am not overly academic, I thought studying only Chinese would probably not be very employable, so I decided to combine it with Management Studies. I applied to Durham University in England and got accepted, but prior to starting I thought I had better visit China to confirm I had made the right choice, so I applied for a gap year, worked some casual jobs and then headed for China.

You seem crazy enough

I arrived in a Beijing under martial law on 26th May 1989. I did not really understand what martial law was and headed out to stay at Peking University. I then went to have a look at Tiananmen Square, which was full of people protesting against government corruption. I went back again on 2nd June to have another look. I climbed onto the Monument to the People's Heroes, which is right in the centre of Tiananmen Square and which was where all the foreign news crews were reporting from. I got a CNN cameraman to take a picture of me with the portrait of Mao on the wall of the Forbidden City in the background and a sea of protesters in between. The atmosphere was lethargic and the whole square smelled like the elephant house in the zoo due to the many protesters living in the square, the heat and the poor sanitary facilities. I left and biked back to the university.

In Tiananmen square a few days before the June 4th 1989 Incident. Picture taken by a CNN camera man from on top of the Monument to the People's Heroes. The Forbidden City and Mao's portrait can be seen in the background.

The night before June 4th heavy thunder could be heard in the distance. Little did I know that it was not all thunder and the next day, as I made my way to the front gate, I saw students making Molotov cocktails, looking worried, and barricading the front gate in anticipation of the arrival of the People's Liberation Army. At this time most foreign students at the university were calling their respective embassies to be evacuated and embassy cars had started arriving to pick up students.

I called my parents and told them that this was not exactly what I had expected from a holiday in China. They had not yet heard the news so they said the whole thing sounded quite exciting. Not wanting to worry them I did not explain what had happened.

I then called the Danish Embassy, which was quite surprised to get a call from a tourist staying at Peking University and it was arranged for a car to pick me up the following morning. It took forever for the car to arrive and on the way back to the embassy I found out why. Burnt-out buses and army vehicles were blocking the road. Eventually we made it back and I was put up with one of the diplomats. I decided to call my parents again to tell them I was staying at the Danish Embassy. The phone did not even ring once before my dad answered despite it being **4AM** in Denmark! June 4th had now made the news and they were pleased to hear that I was in a safe place.

I was not keen to go back to Denmark on the official evacuation plane, so I called Thai Airlines to see if I could change my ticket and fly to Thailand instead. Yes, they did have places on a plane, but I would have to come downtown to have my ticket reissued. I jumped in a taxi and drove towards the centre of town. As we were nearing the second ring road I could see eight tanks parked on the flyover and soldiers lying barricaded behind cement blocks in front of the tanks. A lot of guns were pointing in my direction and the taxi driver refused to go any further. *They must be able to see I have yellow hair,* I reasoned. The Chinese call blond hair 'yellow' and it is very handy to have, because this precludes you from being mistaken for a local person, who all have black hair. As June 4th was a domestic matter this was a factor working to my advantage. I got out of the taxi and walked towards the Thai Airlines office and the roadblock. Halfway there the tanks started firing. Usain Bolt would have been proud to run at the speed with which I entered the Thai Airline's office! I managed to change my ticket and by the time I got out again the shooting had stopped.

UPDATE

China continues to have an uneasy relationship with the June 4th Incident. When the English version of this book went on sale in China in 2003, we had to cut out the above photograph of me in Tiananmen Square because it was deemed too sensitive. However, we did not have to cut out the text related to the June 4th Incident. This seemed strange, but on reflection it made sense. The book was only available at expat friendly outlets, which were rarely visited by local Chinese because it was still too expensive for them to shop there. Moreover, at US$20 my book was outrageously priced for a local Chinese who was used to pay, at a stretch, US$3 for a book. So the risk that a local Chinese bought my book and read it, in English, was remote, yet it was conceivable that a generic Chinese with limited English ability, would wander into an expat outlet and browse through my book before putting it back on the shelf. Hence the censorship of the picture, but not the text.

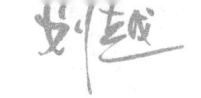

When the Chinese version of my book was published in 2004 all mentioning of the June 4th Incident was omitted. This I fully expected, but there were other things about translating my book into Chinese and publishing it in China, which surprised me. The translation of the manuscript was complete rubbish, but my publisher Shanghai People's Publishing in cooperation with Bertelsmann Direct Group told me that was normal. Foreign books were not sold on the quality of the translation, but rather on the quality of the design so the Chinese edition had to be full of pictures, or no one would buy it. I guess the publishing industry had become warped like that because there were not enough qualified translators to go around and those who were qualified were too expensive to engage for a book full of photographs that would retail at up to US$3. I liked the idea of having a beautifully designed book, but I could not accept that we would publish translation gibberish, so in the end I paid for the translation myself. The Chinese edition sold 20,000 copies, and could potentially have sold more, had the cooperation between Shanghai People's Publishing and Bertelsmann Direct Group succeeded.

The whole experience made my resolve to study Chinese stronger and when I returned to Beijing in 1990 after one year of studying at Durham University things seemed back to normal, except for the odd bit of melted metal on the streets, serving as reminders of where the busses and army vehicles had burnt. I studied one year at Renmin University and then went back to Durham to finish my degree. After graduating I returned to Beijing to work and it was during this time that I had a funny experience, which became a key influence on my rowing project.

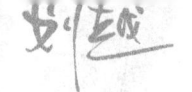

Yang Boning, my former classmate from Atlantic College, was now teaching English at university and he invited me to come and give a talk about Atlantic College. So one early winter morning I found myself standing on a podium looking out on 500 Chinese students packed tightly into the lecture hall. For the next hour I told them all about Atlantic College, international understanding, community services, and the fantastic experience it had all been.

'Any questions?' A young girl in the front row put her hand up. 'Yes,' I encouraged, relieved that at least one person had understood what I had been talking about. She stood up.

'I recently read in the newspaper that a three-year-old boy got into an airplane on the west coast of America and flew it alone to the east coast where he landed safely. This I will call nothing less than a miracle. Do you think the Chinese government will support similar initiatives?'

Yang Boning was sitting behind me and I turned to him with an 'is she for real' look on my face. He smiled, urging me to reply. I wracked my brain and finally came up with a politically-correct answer.

'It is not very responsible for a three-year-old to fly a plane by himself and I therefore do not think the Chinese government will support similar initiatives.'

She looked disappointed and there were no more questions. The session finished shortly afterwards.

I went back to my office wondering where the hell that question had come from. It had absolutely nothing to do with what I was talking about. On reflection, it slowly dawned on me.

Back in 1993 information flow in China was pretty limited. People were therefore looking for any scrap of news they could find and because they did not have much to evaluate it against, they were prepared to believe anything. For all this young girl knew, it could be perfectly normal for a three-year-old American kid to fly around in his own plane. After all, America was a capitalist country! She had wanted me to help explain the world outside and I had failed miserably!

At the same time it also showed something else. The average Chinese has a huge appetite for weird and wonderful adventures, the weirder the better, because it helps them expand their horizons and push back borders. This was reconfirmed to me in 2000 when Zhang Jian became a national hero overnight after swimming 123 kilometres in 50 hours across the Bohai Strait in northern China. Few Chinese had thought that possible. For most Chinese swimming is still an Olympic discipline, which only takes place in an indoor swimming pool.

National hero Zhang Jian about to commence his 123km swim across the Bohai Strait. He had of course chosen an auspicious start date with lucky numbers: 8 day of 8 month, 2000. "8" is considered a lucky number in China because it sounds like "prosperity", so no wonder that the opening ceremony for the 2008 Beijing Olympics was held on 8th day of the 8th month at 8PM. The closing ceremony was on the 24th of August, which is a less lucky date, since it has a "4" in it. "4" in Chinese sounds like "death", so this number is avoided like the plague. But despite this, the 8s outweighed the 4s and China won 48 gold medals, 12 more than the USA who came in second place with 36 gold medals.

* * * *

All this I was recalling as I was sitting in the bar on Koh Samet, looking at the glowing sunset and sipping a Singha beer. The rowing project was coming together and I thought to myself: *If I build the boat in China and row with a Mainland Chinese it would be interesting for both Westerners and Chinese. Everyone will be able to identify with one of the rowers and, while no one will believe the project is possible, at the same time it will be interesting to follow from an armchair because of the adventure element. And when we do succeed, we will have helped to expand people's horizons.*

Not bad, I thought. I ordered another beer and continued my internal dialogue: *I will then superimpose the ideals of the United World Colleges on the project and showcase international understanding by demonstrating that it does not matter how different the cultures are and how seemingly impossible the task, as long as both sides are willing to cooperate and trust, everything is possible! That could be a good lesson for all the struggling joint ventures in China.*

While at university I had done my dissertation with Durham University Business School. I had focused on 'Working on site', i.e. what happens to Sino-foreign projects after the deal has been struck and it moves from a planning to an implementation stage. I had had great fun researching my dissertation by visiting remote building sites around China where foreign engineers were struggling to get things done. My dissertation received the highest mark Durham University Business School had ever awarded to an undergraduate. Its key findings were that working cross-culturally, across living standards and across value sets is very difficult and prone to significant misunderstandings. It also outlined a number of ways to deal with this, such as ensuring there is a joint vision and sufficient humour to deal with crises. However, these problems are not solved overnight and I therefore thought many joint ventures in China could still benefit from some positive inspiration about working cross-culturally to achieve seemingly impossible tasks.

UPDATE

Sino-Foreign Joint Ventures (JVs) used to be the only way for foreign companies to establish a foothold in China, apart from using an agent. After China became a member of WTO in 2001 it became possible to register a wholly foreign owned enterprise (WFOE) in China. Top Chinese talent used to prefer working for foreign companies in

China, but as China has opened up and developed its own global brands, China's most talented workers have started to prefer working for Chinese companies as they offer better career opportunities, at comparable or better pay. Foreigners are also increasingly working for Chinese companies in China thereby willingly choosing to become the cultural outsider in the workplace. Moreover, Chinese enterprises are acquiring companies outside of China. As a result, many individuals around the world, who have never had any previous interest in China or the Chinese, suddenly find themselves being cultural outsiders working in a Chinese company for a Chinese boss in their home country. What a shock to the system this must be, because the Chinese of course want to shape the corporate culture on Chinese values, just like a Western company in China would want to shape its corporate identity on Western values. The Westerner and the Chinese are thereby equally ethnocentric, but the big difference is that the Westerner is historically unaccustomed to being the weaker party in the equation, which is hard on the ego. This struggle about which culture should be used as the baseline for cross-cultural cooperation and understanding is likely to intensify with China's increasing confidence, internationalism, and power. A new paradigm for international understanding will have to morph into existence. Bumpy ride ahead...

The project was getting better by the minute and after a while I had a further brainwave: *I will use the project to fund scholarships for Mainland Chinese students to go to United World Colleges. In that way I can give something back to UWC and help to make China better represented there. Raising scholarships will also help to introduce the concept of combining sports with charity to the Chinese who mainly participate in races to win prize money.*

The project had now taken shape. It would combine living a personal dream with helping others and promoting international understanding.

The last thing I thought about before I fell asleep was whether I could find a Chinese rower. When I had lived in China in the early 90s I had not met anyone who seemed to have the physical or mental potential to do the trip. I thought that the physical aspect was simply a matter of training. What would make or break the trip was mental stamina.

During World War II, Atlantic College co-founder Kurt Hahn was asked by the British navy to investigate why so many of its younger sailors died when shipwrecked whereas, seemingly illogically, the older and less physically-fit sailors survived. Hahn concluded that young sailors perished because they thought they had reached their limit sooner than the old sailors and subsequently resigned themselves to dying.

Hahn established a training programme for the sailors, with the mission of pushing the young sailors outside their comfort zones in the great outdoors. At the end of these experiences the young sailors would then have gained confidence, pushed back their limits, and would be tougher and mentally fitter for survival. The concept worked! After the war it was adopted for civilian use and Outward Bound was founded, and Hahn also incorporated the concept into the community services at Atlantic College. They help students gain self-confidence.

UPDATE

> Atlantic College used to be a recognised call-out station for Inshore Lifeboats, Cliff Recue and Beach Rescue, where students would brave all weather conditions to save members of the public out at sea and along the rocky coastline. These services were set up by Atlantic College's founding fathers and involved a certain amount of danger, so the students had to know what they were doing. Students would train during their first year and obtain qualifications allowing them to run real rescue missions during their second year. Second years were also responsible for training their first years. These

rescue community services saved over 150 lives over the years. Unfortunately, during the 2010s these services were shut down one by one due to increasingly stringent Health and Safety legislation. As a result, the seafront at Atlantic College, which used to be buzzing with activity, is now deserted and Atlantic College has lost its original core community service program. One glimmer of hope for the revival of the seafront is the recent AtlanticPacific initiative by alumnus Robin Jenkins to carry on the proud student boat-building tradition at the sea front. Students at Atlantic College invented the R.I.B. (Rigid Inflatable Boat) and, in the spirit of community service, sold the patent to the RNLI (Royal National Lifeboat Institution) for £1 in 1973, enabling the RNLI to make the R.I.B. into the global success it is today. Atlantic College never cashed the £1 cheque.

The young girl concerned about flying and most other Chinese were clearly lacking the exposure to have built sufficient mental stamina, but recently the Royal Hong Kong Yacht Club had donated a number of old Etchell racing sloops to the China Yachting Association. China's national sailors had come down to Hong Kong and together we had sailed the Etchells back to China. I had been impressed with both the physical and mental attitude of these sailors and I felt pretty confident I could now find a rower who could go the distance.

This is a great project. It deserves to succeed, I thought enthusiastically, drank up my beer, paid my bill and headed for bed. The sun had long since set.

When I woke up the next morning and the effect of the beer had worn off, I still thought it was a great project, but it now appeared a lot more complicated than the night before! I headed for the airport and flew back to Hong Kong.

Taking the plunge

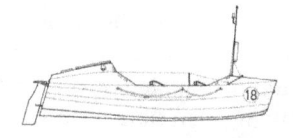

It is one thing to sit on a tropical beach under the influence of alcohol and dream up crazy ideas. It is something else to come back home and announce it to your friends and colleagues. It brings a new perspective to things.

Would they think I was a complete fool and that I had lost my marbles? Would they even believe me? Back in Hong Kong I called Rob Stoneley, a friend of mine from university who happens to be a rower.

'Hi Christian, I guess you are back from your sailing trip to Singapore. How was it?'

'Really boring, there was no wind. But I signed up for a rowing race across the Atlantic in 2001. I am going to row with a Chinese and raise scholarships for sending Mainland Chinese students to a United World College!'

'You are *mad!*'

This was comforting. Rob did not think I was an idiot, nor that I was sure to die, or that I was kidding. He simply thought I was mad. Given the news I had just given him I thought this was a pretty positive endorsement.

I then told my best friend Rory Forbes from university who is now a police inspector in Hong Kong. He responded by giving me a big pot of petroleum jelly:

'You will get a sore ass. It will be pretty tough not to have sex for that long. Try not to use it all making out with your rowing partner.'

Another stellar endorsement!

But I could not bring myself to tell my mother. She is used to me doing pretty strange things, but I had a feeling this would be off the scale.

I did not tell anyone at work either. I carried on working as a consultant and paid my quarterly race fees to the Challenge Business,

Taking the plunge

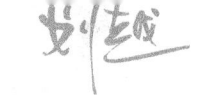

but in early 2000 I was getting increasingly jumpy. I either had to give up the idea of doing the row or I really had to get my skates on. I decided to announce it at work. It got a mixed reception, but I don't think anyone took it seriously. A colleague of mine gave me a copy of Rob Hamill's book *The Naked Rower* with a note saying 'To the guy who does not have both oars in the water'. Rob's book told the story of how he had made it to the starting line and across in the 1997 race and it soon became the reference Bible for me over the next year.

It was understandable that work was not too interested. Working 70 hours a week as a management consultant, often on weekends, does not leave time for preparing to row the Atlantic. I knew that and work knew that. I could not afford to give up working as I needed the money to fund my campaign and also to be allowed to stay in Hong Kong, as work sponsored my visa. I knew that and work knew that.

Then a client approached me to ask if I wanted to come and work for them. I had done good work for them already and I knew they liked me, but changing jobs would not solve the issue, so I said no thank you. The client persisted and eventually I said:

'OK, but only as an independent consultant and only three weeks per month. And I need to finish at the end of July 2001, because I am going to row across the Atlantic.'

'Give us a proposal,' the reply came back.

I did, they accepted. I incorporated a company in Hong Kong to be able to sponsor my own work visa and resigned from my previous job in August 2000.

By luck my visa and financial worries were thereby solved and I was now a business owner, even if I only employed myself. For the next year I worked less, earned more, and had a great time travelling around the Asian operations of a global insurance company implementing a financial model I had written and training staff on product and distribution channel profitability.

Things were starting to pan out and my change of job signalled to friends that I was really serious about the row. Coupled with changing jobs I also decided to make changes to my lifestyle. Hong Kong is fast paced with the stereotypical expatriate being preoccupied with how much money (s)he makes, when (s)he will get promoted, how drunk (s)he got last night, and whether (s)he has had sex recently. I pretty much fitted the profile, so major changes were called for in addition to the recent change of job.

Giving up drinking was easy. After Koh Samet I had soon afterwards been on another diving trip to Sipadan, an atoll in Malaysia with fantastic marine life, including manta rays and hammerhead sharks. Twenty metres from the beach the bottom of the sea falls away in a several-hundred-metre vertical drop. I had been very careful diving there but, nevertheless, I returned to Hong Kong feeling a funny tingling in my body. The next day at work, as I was sitting in a meeting I suddenly lost vision in my right eye, signalling it was time to find a diving doctor and get recompressed. After several recompression trips my vision returned, and the tingling on the right side of my face subsided, but my right arm was still funny. However, there was nothing more the doctor could do except assure me it would gradually get better over time. It did, but every time I drank alcohol the tingling came back. It was therefore easy to stop drinking alcohol and suddenly I got up much earlier on the weekend and could get a lot of training done.

Although I had been very fit until I left Atlantic College, I had hardly done any exercise in the 12 years since then. I really needed to kick start my fitness campaign. The annual Hong Kong 100 km MacLehose mountain run in support of Oxfam sounded like a good challenge. I asked around and eventually managed to join a team of 'Super Trailwalkers', people who had done the race before in less than 20 hours. The team consisted of Rupert, Malcolm and Alex. Rupert and Malcolm were accountants and Alex a heavily-tattooed sales manager. We started training about six months before the race start. Initially I could not run more than a few hundred metres in the mountains before I had to stop. It must have annoyed the hell out of my teammates, but particularly Alex was good at promoting the team spirit and they would wait for me. We continued training like that, always sticking together, and after a while I was not the slowest runner any more. We eventually completed the race in 18 hours and 41 minutes, which was a personal best for all of us. The main reason for the result was good team dynamics. You do not stay on top all the time for such a long race, but there is always one person on top who then leads his teammates until he fades and another teammate suddenly gets a second wind and takes over driving the team. Another thing I learned from the race was the importance of having the right equipment. I always thought that flashy equipment is unnecessary, but when you are really pushing yourself the right equipment is key both for protection and for comfort. Finally, it gave me internal confidence that I could get my body back into shape for the task of rowing across the Atlantic.

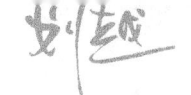

About the same time as I set up my own company I started a relationship with Véronique, a attractive French girl who did not seem to think I was too crazy.

We took a holiday in August between me changing jobs and went to London to call in at the International Office for United World College to present my project. This could really put UWC on the map in China and if we all worked together we could raise a lot for scholarships. I had pictured that we would all be high-fiving out of excitement when I told them of my plan and that we would immediately call Nelson Mandela (who was then President of United World Colleges together with HM Queen Noor) on his mobile phone to get his support, but reality was much different.

Elaine Steel, the Director General, was not enthusiastic. They had supported similar projects before, but in the end the ex-students just wanted to leverage the United World College name to help raise funds for their own adventures, with nothing in it for UWC. I tried to impress on her that I was serious about raising funds, but she remained unimpressed. After three hours of intense negotiations, she thought she had dealt me a killer blow:

'If we were to support you, you must pay for all the race costs yourself so that every penny raised goes towards scholarships.'

This was quite a steep demand. She was basically telling me that I would have to pay US$90,000 for the privilege of obtaining an endorsement to fundraise for UWC. Mainly out of spite I agreed. I could afford the money thanks to my new work contract and I just could not accept that she would not see the potential benefits. If US$90,000 would solve the problem, then so be it, because I was confident that if the UWC PR machine got behind the project we could raise US$90,000 in scholarships many times over.

The reply surprised her and she told me to write a project proposal that she could put to the board in November for approval. I tried to get her to pre-sell the project to the board members since a decision in November would leave less than one year to undertake the fundraising before the race started the following October, but this was not possible.

The only slight help I got was she promised to write to Thor Heyerdahl to inquire if he would be the patron for my trip, as he was a keen supporter of UWC. In 1971 he had brought two ex-students from Atlantic College on his Ra expedition across the Atlantic, sailing the papyrus boat *Tigris*, and I thought my rowing trip and fund raising cause would therefore appeal to him.

I left the International Office feeling depressed. It was the first time I had dealt with the International Office and it was in stark contrast to the Atlantic College campus where everyone is very dynamic and committed to action. In defence of the International Office I have since learned that it is generally difficult to obtain endorsement from charities to legitimise fundraising for them. They would much rather just receive the cash afterwards. Providing endorsement incurs risk. What if the people fundraising are crooks and run off with the money? What if they die in their pursuits? That could make a charity look bad. Granted, but as an ex-student I felt I should be given more credit than that.

UPDATE

> In 2016 the International Office launched a new cooperation framework to accommodate initiatives that want to use UWC branding to support and promote UWC. This set up is called UWCx. So the problems we experienced should now be a thing of the past.
> See https://www.uwc.org/uwcx for details.

Véronique and I arranged to visit the Challenge Business in Plymouth to see one of the rowing boats. The trip did not get off to a good start. We were having breakfast at my cousin Edward's place south of London when he asked: 'Where are you going today?'

'Just down the road to Plymouth,' I replied.

'Plymouth is not down the road. Plymouth is in Devon very far from here!'

We jumped up and dashed to the car, got in and drove like crazy, and luckily arrived in Plymouth on time. I was annoyed with myself. How was I going to navigate across the Atlantic, when I could not even tell Plymouth from Portsmouth?

The Challenge Business' office was situated on a navy base so we had to be met at the entrance. I first thought it was Teresa's assistant who had come to greet us because she looked so young. From dealing with the very helpful and professional Teresa on the phone I had a mental picture of an older person. But Teresa it was and she took

Taking the plunge

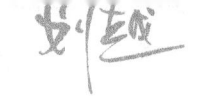

us to see one of the boats. It was exactly as I expected, which made me feel good.

'Are you one of the nutters doing the rowing race?' asked one of the staff. I answered in the affirmative. He continued: 'It's a great race — just a shame about the high mortality rate,' and walked off.

I thought nothing of it, but I could see that Véronique was not impressed. 'Too close to the truth,' she seemed to think, gazing at the seven-metre-long fragile-looking rowing boat.

The rest of the visit went well. I asked Teresa whether I could have race number '8' because it is a lucky number in Chinese as it sounds like 'rich' and would therefore help my fundraising campaign. Unfortunately, it was taken, but '18' was available so I settled for that.

'The more luck the better!' I thought, feeling very happy with number 18. It sounded like 'definitely rich'.

At the end of the holiday I went back to Hong Kong. Véronique stayed in Europe. Before we started going out she had already planned to leave Hong Kong and arranged a job transfer to Germany. Driving back from Plymouth we discussed our situation and decided that there was no point in her coming back to Hong Kong until after the race, as preparations would consume all my spare time. In addition, during my three weeks of work per month I was always travelling so we would not be able to see each other. We settled for a long-distance relationship.

Back in Hong Kong I started looking around for somewhere to build my boat. Through the Royal Hong Kong Yacht Club I got in contact with Keith Mowser, who had a joint venture in Southern China building 470 and Laser dinghies as well as high performance kayaks. He invited me to present my project plan to his joint venture partner, Mr Guo.

One early morning we met in Central and caught the train to Lo Wu where we queued for hours to get across the border. We then battled our way through the crowds to the bus station where we got on the local bus to Shanwei. Two hours later we got off the bus and were met by the company driver, who, after driving through the small village of Hong Hai dodging domestic farm animals and down a very poorly maintained dirt road, dropped us off safely at the Luyang Boat Building Company factory.

From there on everything happened in Chinese. We climbed the stairs to the third floor and entered Mr Guo's office. He looked to be in his mid-forties and with his beaming smile I could not help but like him. Keith introduced us and we made the normal pleasantries, including the exchange of business cards, which is a must in China. We then went next door to the conference room where the management team was waiting. The managers of woodwork, fibreglassing, metals, and the foreman looked at me expectantly as I started my presentation. At the end of my explanation of why I thought they should build my boat, the look of expectancy had changed to disbelief. We shuffled off to the staff room where I showed them the 1997 race video on their impressive karaoke system and the attitude changed again. They could see that the boats and the race really did exist. Mr Guo later said in an interview with CCTV: 'At first I thought Christian was crazy, but then I saw the race video and that helped.'

We walked back into Mr Guo's office and sat down. He gave me his proposal: 'The race is very interesting and we would be proud to build the boat. Imagine our company building the first Asian-built ocean rowing boat, which will be rowed by the first ever Asian to row across an ocean! We cannot lose our shirts on this, but we are prepared to build it at cost.'

During my presentation I had suggested he should sponsor building the hull because of the uniqueness of the project and because I expected we would be able to drum up a lot of media interest, which would also reflect positively on his company. I also really wanted to do the whole project on goodwill, instead of based on money, just to show that things can be done differently in China.

When working in China as a foreigner you are frequently met with knockbacks like 'this is not how things are done in China' and 'this is not mutually beneficial'. Put very black and white, this basically means the Chinese side thinks you should pay more or they should have things for free. In the early 1990s while working in Beijing, I employed a sales engineer to sell imported pharmaceutical packaging machines to the Chinese industry. One day, on the way to a client visit, she told me that prior to working for me she had worked for a design institute. They had imported a tablet pressing machine from Germany, taken it to bits, reverse-engineered it, and sold the blueprints to factories around China. She proudly announced that as a result it was now hard for the German company to sell its machines in China because they were too expensive and that all spare parts, even for the real machines, were now pirated copies. I told her that, to me, it did not seem very fair to the German company, but she replied that it was mutually beneficial because that machine was now the market leader. It did not seem to matter to her that it was the pirated machine that was the market leader. The German company should be happy anyway. It was an endorsement of their product!

When we were looking at publishing this book in Chinese I spoke to some people in the industry about how many copies a book would have to sell to become a bestseller. 'It is not measured in books sold because if a book is popular it will be copied straight away and we cannot track the pirated copies. Instead we gauge a bestseller by how long the press talks about a book. Anything more than two weeks is deemed a bestseller,' they replied.

This is a real paradox of China. Having your product copied really is a sign of success, you just happen to lose out on the revenues!

Foreigners operating successfully in China have developed a fine feel for whether 'this is not how things are done in China' objections are real or whether the Chinese are trying to pull a fast one. It is understandable they try to squeeze you. China is after all a developing country and, coming from a developed country, chances are you have more cash than they have so it would not be so bad if you paid over the odds. At least it used to be like that, but China has come a long way in the past 10 years. A university friend of mine, Rupert Hoogewerf, now publishes the 'Hurun Report', estimating the wealth of China's top entrepreneurs. One look at that list will show that some Chinese are now incredibly rich by any standard. So the 'this is not how things are done in China' comment is increasingly simply a negotiation tactics.

UPDATE

> China started taking IP rights more seriously after Chinese companies also started copying Chinese companies and not just foreign companies. This shows two trends. Foreign brands are not necessarily the most coveted brands in China anymore, and increased domestic economic development and prosperity calls for better laws to regulate the market.

With my rowing project I also wanted to test how far China had developed since I left in 1994. Would China now be so far ahead that I could find a Chinese company that would be willing to sponsor building the hull? I looked across the table at Mr Guo and evaluated his offer. His offer was very progressive for a Chinese, but I thought it was too slanted in his favour. If we succeeded, he would ride a huge PR wave for free. If not, he would not have lost anything. 'That is a good offer, but I am really looking for a partner who is willing to invest in building the hull against the return of media exposure, which will be huge when we have succeeded. Imagine the first Chinese to row across the Atlantic Ocean! That will even outdo Zhang Jian's swimming achievement!'

To me it was obvious that we were going to succeed, but in hindsight I can understand Mr Guo's reservations. I had never rowed before, the concept of rowing the Atlantic is difficult for anyone to fathom, and he knew better than me how difficult it was going to be to find a Chinese rowing partner. I was walking around in blissful ignorance!

I went back to Hong Kong without agreeing with Mr Guo to build the boat. There I had a re-think. What to do now? I reasoned that if I got some press coverage then I had a better selling point when I contacted potential sponsors, but I could not find any journalist who was interested in writing about my project.

'Maybe your story sounds a bit too fantastic,' Inge Strompf-Jepsen, a fellow Danish member of the Yacht Club, told me over lunch one day. 'Why don't you get that boat built? It will make the whole thing more real and show your commitment.'

I knew she was right, but I really did not want to accept Mr Guo's offer. There had to be something better. I therefore settled on the next best thing. I decided to build a website to try to drum up interest. I thought it would sound good to call the media and say 'Why don't you check out my website!' It was the height of the dotcom bubble and websites were all the rage.

I went home and started searching for a suitable domain name, which would work well in English and Chinese. I settled on 'Duhai', which means 'crossing the ocean'. The Chinese characters are aesthetically pleasing to look at as they both have a water radical on the left side. 'Perfect,' I thought and called Yang Boning in Beijing for a sense check.

'I don't think that will work Christian. We use that term when we talk tough about Taiwan. "Duhai" is a synonym for China invading Taiwan.'

'Probably not the right name for a project which aims to promote international understanding,' I observed and hung up.

Taking the plunge

I got back on the computer and continued looking. Finally I settled on 'Yantu', which means 'in a state of motion' or 'underway' and after checking with Yang Boning I bought Yantu.com. I liked the name. It sounded a bit like Yahoo, which was worth truckloads of dollars. A new idea had dawned on me. Maybe I could finance my rowing trip on a dotcom business model. It was September 2000 and startups like China.com, Sina.com and a host of other dotcoms with more hare-brained ideas than mine were attracting large amounts of investment. I sat back and enjoyed my imaginary wealth for a while before I got back to reality.

Through friends in Hong Kong I got in contact with a guy in Australia who could code the site. I started writing content and through another friend, Inge Nielsen in Beijing, I got hold of Sun Hongmei to translate the content into Chinese. I wanted to make a bilingual site. In true Internet-age fashion I ended up with a web designer in Australia and a translator in Beijing while I was living in Hong Kong!

Despite feeling confident I was about to become an Internet billionaire (based on future projected earnings, and subject to raising seed capital) the fact that I did not have a rowing boat was weighing heavily on me. One September afternoon I met Keith Mowser by chance in the bar at the Yacht Club.

'How are you getting on with the boat?' he asked.

'I am getting nowhere,' I replied, looking into my lime-soda, my favourite drink now I had stopped drinking beer.

'I have a 40-foot container about to leave from the UK for here. We can put your boat kit in there for free. Why don't we build you that boat?' he offered. I had not been able to find any other alternative and I was sure they would do a good job, so I finally swallowed my pride.

'OK, but at cost. Luyang will not be allowed any branding on the boat or our clothes, but you can brand the oars,' I said.

The deal was struck and I ordered the boat kit from the Challenge Business in the UK. The race was a one-design race meaning we would compete in identical boats. This was done to ensure that everyone went to sea in a seaworthy boat, and also to ensure a fairer race as it precluded teams with more money from gaining competitive advantage through using a faster design or lighter materials. The Challenge Business had commissioned the design from the well-known designers Peter Rowsell and Phil Morrison. The kit was the same as that used in the 1997 race, except for minor adjustments to the skeg (part of the keel) and the addition of a spray rail.

Three weeks later the kit, comprising 25 sheets of flat-packed, laser-cut marine plywood, arrived in Hong Kong. Keith called me with the news.

'It might be cheaper to smuggle the kit into China on a fishing boat, but given that you need to get it out again, we better go through the official channels,' he said. A truck was arranged.

Keith had a point. How difficult would it be to get *Yantu*, as I had decided to name my rowing boat, back out of China? This could be an adventure in its own right and we would not know until we tried.

At the end of October I went to the factory. A cover had been arranged in the courtyard and a number of workers were milling around the boat, putting it together. The whole setup looked very primitive, but then again it was supposed to be a DIY design. The kit was self-jigging and the manual said it could be try-assembled in four hours. 'We did it in three,' a proud Mr Guo exclaimed. The assembly had not been without drama. Keith had warned he would come down like a ton of bricks on anyone who took a saw anywhere near the kit and he and his Hong Kong business partner, Mr Kwok, had been present for most of the assembly.

Basic facilities and great workmanship. 'Yantu' starts to take shape.

The kit was now ready to be glued together with epoxy and we discussed how to proceed. Too much epoxy would result in a heavy and slow boat. Too little and the boat might not be strong enough and break apart. Not for the last time we pulled out Rob Hamill's book to see what he had done. He had opted for fibreglass strips only at the stress points and apart from that gone for an assembly as light as possible. We decided on the same and Keith and Mr Kwok, both seasoned boat builders, began giving instructions. Things looked under control. Before heading back to Hong Kong we had dinner with the local mayor and customs officials. It could never hurt to be on good terms with them. The mayor was very excited that the first Asian ocean rowing boat was being built in his village! A good sign.

Back in Hong Kong I thought things were now starting to move along. Seeing the boat being built was something tangible, both to myself and to anyone else. I also published my first article about The Yantu Project in the Royal Hong Kong Yacht Club Magazine, *Ahoy!*, thereby taking my project public. It was all happening. I had taken the plunge!

Feeling upbeat I decided to tackle the next problem — finding a rowing partner. Véronique was visiting and in early November we flew to Beijing to get things under way. In my suitcase I had a project proposal in Chinese, which could impress the socks off any bureaucrat!

In connection with the Etchells delivery from the Royal Hong Kong Yacht Club to China, I had got to know Mr Quanhai Li, Secretary General of the China Yachting Association. I was pretty certain that the office of the China Rowing Association could not be far away from the Yachting Association so we went to look for Mr Li, hoping he could provide an introduction to the Rowing Association. Mr Li was somewhat surprised to see us, but gladly took us down the corridor to introduce us to Mr Liu from the China Rowing Association.

Mr Liu looked with interest at my project proposal and the race video. Part of my proposal read:

The cost of participating in the race is funded by myself personally, and I therefore cannot offer the China Rowing Association any money for assisting me with my project. However, I can offer:

- An opportunity for a rower to become the first Mainland Chinese to row an ocean (more people have climbed Mount Everest than have rowed across an ocean)
- PRC participation in an international extreme sports event, which will attract international media coverage
- Goodwill through being associated with raising scholarship money for Mainland Chinese students to study abroad, which will further enhance the China Rowing Association's image of good corporate citizenship and internationalism
- An exciting event for the Chinese people to follow.

If the China Rowing Association is willing to support the Yantu project under the above constraints then I propose the following immediate activities:

- The Chinese Rowing Association identifies a number of potential rowing partners from around the country. I come to China and spend one week with the potential rowers from which I select one rower and one back-up rower.
- The right person to undertake the crossing with will be an individual who:
    - is willing to put a lot of effort and personal time into the Yantu project and the ideals behind it;
    - is physically fit to take on the challenge;
    - is mentally fit to take on the challenge;
    - has a sense of humour;
    - can control fear and does not panic when the weather gets rough;
    - can handle being out of sight of land for 50-90 days; and
    - I can get on with in cramped and uncomfortable conditions for the duration of the race.

'This will be a really great project,' I explained. 'It will all work on goodwill. There will be no overseas trips or banquets. I will guarantee all the costs so that we can get to the starting line. I will also pay for my rowing partner to come down to Hong Kong for training. What your part will be is to find me a few potential partners and ensure that the chosen partner will have the travel documents and time off required to train in Hong Kong for one week per month until the race and for the race itself.'

'This is pretty exciting,' Mr Liu agreed, 'But we have never done anything like this before. We normally work with an organisation, not an individual, and the race is not officially acknowledged in China. I will have to ask the State General Administration of Sports for permission, but in any case we will do what we can to assist you.'

Véronique and I left the meeting feeling good and went to call on a number of newspapers. The middle-aged woman wearing a thick, patched quilt at the *China Sports Daily* showed us into a small cold room and poured us hot water from a large Thermos flask. Studying my project plan with great concentration she looked very much an old-school official. We were freezing and expecting the worst.

'I see that you will set up a scholarship fund. Can we administer that?' she asked.

'No,' I said, thinking, *here we go....*

'But you want us to write about your project. It costs money to print a newspaper. Where will we get money from to print your story?' she asked.

'This story has great news potential. If you report on it you can sell more newspapers, which will generate more money for you,' I volunteered, giving her a 101 in capitalism.

'That is not how things are done in China,' came the reply and the meeting ended shortly afterwards.

I was furious! Here I was investing heavily to create a project that would be of great news interest to the Chinese public and would give Chinese students the opportunity to study abroad. There was no way that I would pay to get this story in the press! It deserved better. Frustrated, we flew back to Hong Kong and Véronique continued to Germany.

Getting back home that night I was ready to throw in the towel, but at about 11**PM** the phone rang.

'Hello, this is Wang Zengshuan from CCTV Sports News in Beijing,' a voice said in Chinese. 'I have been speaking to Mr Zhao from the China Rowing Association who told me about your project. We would like to report on it. When are you next in Beijing?'

A long distance call from China! *This costs money, so he must be serious*, I thought. Just to make sure, I said: 'Can you please fax me your details and I will look you up when I am back in January.' Shortly afterwards the fax arrived. The light at the end of the tunnel had been switched back on!

But it was soon switched back off. I got an e-mail from the United World College International Office that the board had considered my application. Lord Richard Attenborough had been very excited, but they had not reached a decision because an ex-student working as an astronaut had been worried about the safety aspect of the race, believe it or not! The decision had been delegated to a sub-committee, which was going to meet in February 2001. In the meantime the International Office suggested I should hire a full-time assistant and set up a trust because if the project was endorsed there would be a lot of official communication to take care of.

Starting a fundraising campaign in February, just seven months before the race start, was cutting it fine. I wrote back asking that if I, in addition to paying the US$90,000 race costs, also hired an assistant and incorporated a trust, could then be guaranteed an endorsement in February? The International Office replied that they might be able to make a decision in February, but they could not promise it would be positive. *This is ridiculous,* I thought. I decided to give up on the idea of raising scholarships for all the United World College schools and to focus only on Atlantic College. At least the people there knew me personally. I called my old housemaster, John Lawrenson. He promised to have a word with Malcolm McKenzie, the new principal. A few days later I phoned.

'Hi Malcolm, my name is Christian Havrehed. I....'

'Hi Christian, John told me about you. You want to row across the Atlantic. Sounds like a great idea. How can I help?'

'As part of my project I want to fundraise for Atlantic College in order to give more Mainland Chinese students the opportunity to study there. To make this credible I need a letter of endorsement from you and the right to use your logo.'

'No problem. I think it would be wonderful if you could help more Mainland Chinese students to study here. Right now we have two, but next year we only have funding for one. I will send you an endorsement letter.'

A few days later the letter arrived. Having received the endorsement the problem of a trust still remained. It would hardly be credible to ask sponsors to deposit funds into my personal account. I needed a charity trust. Francine Kwong and Tammy Wan, Hong Kong friends of mine from Atlantic College, and I worked hard to try and find a solution through the UWC network in Hong Kong, but with no success. Although I still had yet to raise a single dollar and few people believed that my adventure would go ahead or that I would live to see the end of it, the local UWC stakeholders were still worried that the Yantu Project would compete for the money they were trying to get for the Li Po Chun United World College in Hong Kong. In the end I asked the Royal Hong Kong Yacht Club if they would be the custodian and they were happy to assist.

Taking the plunge

UNITED WORLD COLLEGE OF THE ATLANTIC
Patron
H M QUEEN ELIZABETH II

President of the United World Colleges
H M QUEEN NOOR
of the Hashemite Kingdom of Jordan

From the Principal
MALCOLM H. McKENZIE

Honorary President
NELSON MANDELA
Former President of South Africa

19th February, 2001

TO WHOM IT MAY CONCERN

I write to commend Christian Havrehed and his Chinese fellow rowers to you. Their adventure, rowing in a race across the Atlantic, is worthy of your support simply for the bravery and initiative that it displays. Allied as it is, however, to the cause of sponsoring more Chinese students to come to Atlantic College, it is irresistible.

Atlantic College has a proud tradition of equipping its students for sailing and for venturing out to sea. Amongst other service activities related to the sea, the College runs the Royal National Lifeboat Station for its 7 miles of rugged coastline. Students at Atlantic College have, over the years, been credited with saving hundreds of lives. It was at Atlantic College that Christian developed his love of sailing, boat building and other aspects of seafaring.

Atlantic College has also welcomed Chinese students for many years. It has been a great privilege to have enrolled some of our very best students from China for over two decades now. These students, like so many from all over the world, come to the College on full scholarships. Recent years have seen a big increase in the difficulty of raising money to fund Chinese students. It would be wonderful to be able to take more from China. At the moment, we have only one Chinese student coming to us in September.

Your support will make a big difference. Please help another young person from China to come to us.

Yours sincerely

Malcolm McKenzie

Malcolm McKenzie
Principal

The letter of endorsement from Atlantic College Principal Malcolm McKenzie was a big help.

The fundraising effort was now legitimate. However, the Yantu Project did not have status as a charity, which would have been tax deductible, but since tax in Hong Kong is only 16 percent, I did not see that as a big problem. I was running out of time. Spending more time to incorporate a charity would not raise any money. Talking to companies might!

Even if the Yantu Project was now less appealing to sponsors because they had to support with after-tax dollars, it was still a workable setup.

I received a fax from the China Rowing Association confirming it was taking action.

> The Chinese Rowing Association has discussed the Cross Atlantic Rowing Race in detail, and officially put it under the authorised plan and submitted it to the State General Administration of Sports. Up till now, we have not received a reply. Once the project is approved, the Association will immediately discuss the cooperation with you in detail.

And there was still the possibility that Thor Heyerdahl would agree to be patron. This would lend extra credibility to the project. Despite our tense relationship, the International Office had written to him and I was waiting for a reply. It came in late January.

> *Dear Elaine Hesse Steel,*
> Thank you for your letter of 15 January 2001 and all the best wishes for the New Year we have already entered.
> I have always been very reluctant at recommending voyages in open boats across the oceans, for the fear that such vessels may fill seas in rough weather and sink. I would prefer a well-built reed boat myself, and to fulfil a promise to Lord Mountbatten I had a Danish and a Norwegian graduate student from United World College of the Atlantic along for five months on the reed ship *Tigris*. I also was a guest onboard a small sailing vessel together with the British Prime Minister and Lord Mountbatten sailing down the Thames. The vessel (without us as guests) was to sail around the world as publicity for UWC. I believe nothing much came out of it, at least not economically.
> Without bad luck, I hope and believe Christian Havrehed will get across safely, but how much that will promote the UWC is a question the International Board of Directors will have to ask themselves. It might of course create some goodwill in China.
> *With best wishes,*
> *Thor Heyerdahl*

That was disappointing, but at least Thor thought I might survive....

Taking the plunge

Thor Heyerdahl brought two Atlantic College alumni on his 1978 Tigris expedition. 'Tigris' is here seen flying the Atlantic College banner.

I got another fax from the China Rowing Association.

> We consider the Yantu Cross Atlantic Rowing Race a very meaningful event, which is instructive to the youth. However, the Ninth China National Games is to be held in November 2001, and the Chinese Rowing Association will therefore be busy with organising the rowing at the National Games. Since the Yantu Cross Atlantic Rowing Race and the National Games will overlap in time and personnel, the Association has no spare capacity to actively cooperate with your rowing race, but we will try our best to offer support and help.
>
> Following is the information of a protégé of Zhang Jian whom we recommend to you, for your reference.

It was bad news that the China Rowing Association would not be able to actively cooperate, but I appreciated their assistance in proposing another candidate. The fact that candidate was a protégé of Zhang Jian was good, since he was China's most famous long distance swimmer and extreme sports athlete, as well as the General Secretary of the Beijing Triathlon Sport Association. His support definitely carried weight. I studied the attached CV in detail. It read:

Sun Haibin CV

Name:	Sun Haibin 孙海滨
Sex:	Male
Birth date:	Nov. 6, 1975
Native place:	Henan Province
Marital status:	Single
Occupation:	College student, third year
Political status:	Member of the Chinese Communist Party
Height:	1.76m
Weight:	70 kg

Sports history:

1990-1992	Trained in biking in Shangqiu, Henan Province. Best result was the champion of the age group.
1993	Joined the army, trained in Unit 81 of the PLA, specialised in long-distance running.

Taking the plunge

1994	9th place in the National Field and Track Championship, 1500m.
1994	12th place in Beijing International Marathon.
1995	Changed his specialisation to iron man in PLA Unit 81.
1995	3rd place in Asian Triathlon Championship, a top player of the champion team.
1997	3rd place in Asian Triathlon Championship, a top player of the champion team.
1997	5th place in Macau Iron Man Championship
1998	Mild Seven Outdoor Quest. Came 6th out of 24 teams. Took part in the two-month training camp for double kayak. With certain rowing ability and played a leading role in the quest. The most capable one among Asian athletes. Helped teammates to finish the four-day endurance event.
Sep-98	Left the Army, entered Beijing Sports University.
1999	Mild Seven Outdoor Quest. Four day endurance team event in China. Came 7th out of 24 teams.

An interesting CV! I was a bit sceptical about Sun Haibin's 'certain rowing ability'. You do not learn to row from kayaking, but at least he had been in a vessel on the water before. He really seemed to be an Iron Man Triathlete and obviously quite good at it, too. This was important as this showed he had mental stamina. The fact that he had spent five years in People's Liberation Army Unit 81 also indicated he was a real achiever. Unit 81 is the elite sports team of the PLA. Its name commemorates the date, August 1, that the PLA was established in 1927.

'Great,' I thought. 'If we get into a fight at sea he will be a trained killer and I will have to improvise!'

But at least I now had a potential candidate, even if he probably could not row and would be able to kill me in a flash. This was a major step forward and in any case I had started to wonder whether an endurance athlete might not have a better mindset for rowing across the Atlantic than a rower used to racing over a few thousand metres.

I called the China Rowing Association and arranged to go to Beijing to meet Sun Haibin. I then called Wang Zengshuan from CCTV and told him I would arrive a few days later. 'Should we book a time now to meet up?' I asked. 'No need, just call me when you get here,' he replied.

Deliberating on the prospect of having an ex-soldier and trained killer as my rowing partner...

Things start to take shape

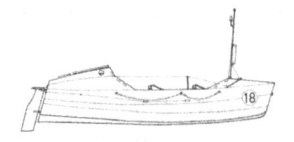

Late january is very cold in Beijing. After dropping my bags at my friend Inge's place, I headed straight to the China Rowing Association to say hello and get the contact details for Zhang Jian, whom I arranged to meet the following day in his office at Beijing Sports University. I called Wang Zengshuan at CCTV to tell him I was in town.

'Hi, it's Christian. I have arrived,' I said cheerfully.

'What, you're already here? I'm so busy I won't have time to meet you. Call me next time you're here,' the reply came.

'Come on!' I said. 'We arranged to meet and you told me I did not need to make an appointment. You said I could just show up.'

'OK, but it will be short. Meet me at the east gate of CCTV tonight at seven o'clock. Goodbye,' a stressed-sounding Wang Zengshuan said and hung up.

What a downer, but at least I had secured a meeting. I went back to Inge's place to get some rest. It was half-past six, bitterly cold and dark when I hailed a taxi to drive to CCTV. I got out at the east gate, which is essentially a badly-lit security gate in the middle of nowhere. Several people were jostling to use the battered phone a guard was handing out through the window from inside his warm office to the cold and waiting masses outside. Eventually I got my turn.

'This is Christian Havrehed. I have a meeting with Wang Zengshuan.'

'He is not here. Try again in five minutes.' Click! The phone went dead.

I stared at the receiver in my hand in disbelief until it was taken away by the next person in line waiting to call someone inside. I joined the back of the queue again, stamping my feet not in frustration, because things like this happen in China, but rather to try and get the circulation going. It was cold!

Eventually, I got back to the front of the queue again and made another call.

'This is Christian Havrehed again. Is Wang Zengshuan back?'

'No, I think he has gone home,' the voice said.

'He can't have gone home. I have a meeting with him. We were supposed to meet at seven o'clock. It is now a quarter to eight. I will wait for him here until eight o'clock. Please try and find him!' I pleaded.

'I will see if I can find him, but I think he has gone home,' the voice replied and hung up.

Gone home! How could anyone be that rude? I was utterly disappointed and extremely cold. Through my white breath I was watching the minute hand on my watch edge closer to the hour. The closer it got, the more worked up I was becoming.

'Fuck the whole thing! If this is how it has to be then I've had it!' I muttered under my breath, feeling increasingly frozen. I made a promise to myself that if Wang Zengshuan was not there by eight o'clock on the dot I would throw in the towel and give up the project. I did not deserve this shit! I had had my fill! And then it happened. With only 20 seconds to spare a friendly-looking guy and an equally friendly-looking women came through the gate, approached me and said:

'You must be Christian. I am Wang Zengshuan. This is my colleague Yu Zhen. She is a reporter. Come, let's go and get something to eat and you can tell us all about your project.' He hailed a taxi and we headed down some dimly lit narrow alleys, or *hutongs*, until we found a small restaurant.

After we had exchanged pleasantries, I had thawed out and we had ordered some food, I told them about the project and the philosophy behind it.

'That sounds really exciting and totally unique! We will definitely want to cover this!' Wang Zengshuan said. 'You say you are meeting Zhang Jian and Sun Haibin tomorrow. I would like to send Yu Zhen and a cameraman along to film that.'

This was certainly coming in from the cold. Now the reporters wanted to film our meeting, which, to me, would be quite a private affair as we would be sounding each other out. But if I turned Wang Zengshuan down, I might have lost the one and only chance I would get. 'OK,' I said, not completely convinced. But this was the best thing

Things start to take shape

I did and signalled the beginning of a great working relationship where we got a five-minute slot on the Sports News every month until the race, replayed the following day on CCTV's international channel. During the course of the project Wang Zengshuan and Yu Zhen became friends of mine.

I called Zhang Jian and Sun Haibin to warn them I would turn up with a TV crew. It did not faze them. Unlike me, they were used to the press.

Zhang Jian's office was located at the end of a long and very cold corridor. It consisted of a desk and a few chairs standing on a cement floor and surrounded by whitewashed walls. Zhang Jian did not look anything like what I thought a long-distance swimmer and national hero should look like. He was short, a little plump and smoked cigarettes. After the normal pleasantries he said:

'The China Rowing Association contacted me because they thought if I can swim across the Bohai Strait then I can also row across the Atlantic. But I am a swimmer, not a rower, and in any case, I am preparing to swim across the English Channel in August. So it is no good. But Sun Haibin, he can do it. He is the right candidate for you.'

I glanced at Sun Haibin who was listening to the conversation intently.

'I understand. But will you be able to arrange for Sun Haibin's travel permits?' I asked thinking I might as well get the sticky points out of the way immediately.

'That is no problem. The China Rowing Association can still help. And I can arrange for him to take leave from university.'

This sounded good and I turned to face Sun Haibin. He had not met many foreigners before and did not speak any English, but was keen to impress.

'I am very fit. Actually I am fitter than you. Look,' he said, grabbing my arm, 'I have strong fingers good for pulling oars!'

I was not so impressed, but he could not be knocked for his enthusiasm and he was smiling all the time from a pleasant and open face. I asked him if he had ever rowed before.

'No!'

'Have you ever been out to sea?'

'No!'

'Do you get seasick?'

'No!'

'How do you know if you have never been out to sea? I get seasick.'

67

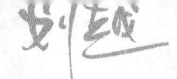

'I don't get seasick!'
'What if we run into a storm in the middle of the Atlantic and the waves get big and nasty. Won't you be scared?'
'No!'
'Why not?'
'I can swim!'

It was pretty obvious Sun Haibin did not have a clue about what it would be like to row across the Atlantic, but he was my only candidate so I decided to give him the benefit of the doubt. And there was something appealing about his personality.

Yu Zhen wanted to get some shots of us exercising together. Outside it was snowing, a far cry from the conditions on the Atlantic where we would expect around 30 degrees Celsius. We decided to go for a run. Sun Haibin suggested we should run through the Yuan Ming Yuan, the old summer palace, burned by the Western imperialist forces in 1860 to punish China for not being more welcoming to foreign traders. He knew the back way in there without paying.

We set off in the snow with CCTV driving next to us filming for a while. Sun Haibin was a good runner. Despite my recent 100 km mountain run it was difficult to keep up. He slowed down and fell into my pace.

'I have been thinking that maybe we need some wrist straps we can sling around the oars to prevent our grip getting too tired when rowing,' he said. My ears pricked up. Maybe Sun Haibin could picture the trip even if he had never been to sea. We left the road and ran in between some trees.

'This is an old execution ground. I normally run here in the mornings before it gets light. Sometimes it can be pretty scary,' he continued. My ears pricked up some more. So he was not all macho. He was willing to open up and show his weaker side, something I considered important, because sooner or later our weaknesses were bound to show and it would be better to know them earlier as opposed to having to deal with them for the first time in the middle of the Atlantic.

'Why do you want to do the race?' I asked.
'I got a teacher to translate your project plan and I watched the race video carefully. Once I saw it, I just knew I wanted to do it! I really want to be the first Chinese to row across the Atlantic!' he replied.

That was not unlike why I had decided to do the race myself and by the end of the run I had warmed to Sun Haibin's personality.

Things start to take shape

We ran back through the main gate of Beijing Sports University, which is one of the few places left in China where you are still greeted by a large statue of Mao. These have been disappearing as Deng Xiao Ping's economic reforms gained pace.

'So, what do you think of Sun Haibin?' Yu Zhen asked, sticking a microphone and a camera in my face.

'He has good potential, but I would still like to look for more candidates as he is the first one I have met. But we have agreed he will come down to Guangdong where I am building the boat to train with me in February for one week,' I replied.

'How well do you expect to do in the race?' she continued.

This was a difficult question. I had done some research on the last race where most teams had arrived within 55 to 65 days. I thought we would probably fit into the top end of that. The race would start on 7th October 2001 and it was Véronique's birthday on December 3rd. To make her birthday we would have to row across in 56 days. That was the answer, I decided.

'The crossing will take 56 days and we will finish among the top 10,' I replied confidently.

'How do you know that?' she probed.

'From looking at how well people did in the last race and besides, if we take any longer, I will miss my girlfriend's 30th birthday and I don't want to do that,' I replied. Yu Zhen laughed.

'You don't look very fit compared to Sun Haibin. Do you think you are strong enough?' she challenged.

Sun Haibin was half a head shorter than me. He had spent most of his sports career doing marathons and long-distance running. Marathon runners do not develop their upper body because they do not want to carry the extra weight. He was therefore lacking upper body muscle mass and I felt slightly annoyed by the question.

'I know Sun Haibin is a much fitter runner than me, but we are rowing, not running, and normally when I start training hard my body gets in shape quickly,' I replied, trying not to offend.

That night CCTV Sports News broadcast a story on the Yantu Project for the first time. After explaining what the project was about, they announced I was looking for a potential rowing partner and provided a contact address.

The programme was watched by around 50 million Chinese viewers with an interest in sport. I was worried that I would not be able to reply to all the applications I was bound to receive, but I could have saved the anxiety. Two people replied!

One was an old friend of mine, Li Jie, whom I had lost touch with. In 1990 he had invited me to join his old middle school dragon-boat team for the annual races on the lake at the Summer Palace. We had made it into the top 10 only losing to the professional PLA teams. However, Li Jie had in the meantime got married and had a daughter, so he was not interested in rowing, but it was nevertheless nice to catch up again.

The second guy was called Lin Miao. I called him to discuss his background. He was into motor sports and was very interested in knowing how big the engine would be on the boat. After uttering significant disbelief that there would not be an engine and that I was talking about *rowing* across the *Atlantic,* he said he definitely could not do that. But he kept in touch and followed the project to the end, which was nice.

'Something has gone wrong,' I thought. 'Surely there are a lot of people who would jump at the chance to row across the Atlantic!' I decided to spread my net a bit wider. The Hong Kong outdoor magazine *Action Asia* made an announcement on their website and I contacted the organisers of the Hong Kong Mountain Marathon Series and Hash House Harriers. I waited in anticipation. After speaking to their members each organisation reported back the same news. No one wanted to row across the Atlantic! I could not believe it! At this point in time a quote from Rob Hamill's book is appropriate. It is a reply he got from one of his potential sponsors.

> ... while rowing across the Atlantic attracts you, it has no similar appeal to others. Certainly from my perspective I cannot think of a more unpleasant way to waste 55-90 days of my life. It would be a lot cheaper to stay at home and thrash yourself from dusk to dawn with a barbed wire whip.

So the talent pool was still one potential partner! After considerable deliberation I decided that Sun Haibin by far showed the most promise. 'The best of the whole lot!' I concluded.

Completing the hull

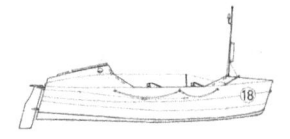

Now that I had secured a potential rowing partner I called Mr Guo to agree on a launch day. Work on Yantu was progressing nicely, and after poring over the calendar to find an auspicious day we decided on Thursday 8th March. It had an '8' in it and it was also Women's International Day, so it had to be lucky. The plan was to row Yantu from the factory down to Hong Kong after the launch, a trip of about 100 miles, and complete kitting her out there.

Having decided on the launch day, a new sense of urgency set in. At the same time the tasks were getting increasingly complicated. The builders had been gluing the plywood pieces together with epoxy, now decisions had to be made. Keith called me to ask what compartments I wanted foam-filled. It is good practice to fill hollow compartments that have no storage purpose with a light-weight foam so that, in case the hull suffers damage and takes in water, buoyancy is maintained because the foam prevents water from flowing in. 'And what about the towing eye, hatches under the rowing seats, rowlocks, riggers, support for the seats, and colour of the boat,' he continued. I did not have a clue and told him I would get back to him. Before he hung up Keith told me he was having problems getting hold of the three Lewmar hatches, which would seal the forward storage room and the aft cabin for sleeping. He had tried everywhere, but all 30, 40 and 60 inch hatches seemed to have been sucked up.

In the interest of speed I decided to call the recommended boat builder for the race. I was sitting in a hotel room in Jakarta where I was working that month. Telephone connections in Indonesia were not particularly good, but I managed to get through to David Graham at Stanley & Thomas Boat Builders.

'Hi, this is Christian Havrehed. I am calling from Indonesia. I am doing the Atlantic Rowing Race and the Challenge Business told me you are the recommended builder. I am building my boat in China and I wonder if you would be kind enough to provide me with some information. What compartments do you suggest to fill with foam?'

'Why the hell are you building your boat in China?' David asked.

'Long story. Can you help?' I pleaded.

'Can you please send me an e-mail and I will see what I can do,' he replied.

'I really need an answer urgently. Can you not tell me over the phone?' I persisted.

David reluctantly agreed. At either end of the phone line we got out the kit assembly drawing, turned to page 33 and, like playing Battleships, David said: 'T4, L1, L6, T9, L4 ...' while I pencilled in the compartments on my drawing.

'Great, thanks. Do you know anything about oars?' I asked.

'You're pushing your luck, mate. We're not even building your boat!'

'Please!' I persisted.

I managed to beg my way through my list of questions. I thanked profusely and hung up. Done, I put it on a fax to Keith. The oars should be 11 feet long with Mecon blades and the shaft should be reinforced in the stress area where it would sit in the rowlock, using five layers of carbon fibre and slimming down to three outside the stress areas. For the two rectangular hatches under the rowing seat we would use the wooden covers from the kit. The seats would be mounted on two eight-inch high hardwood planks running the length of the cockpit on either side of the rectangular hatches and 12 inches into the well. On top of the planks we should use standard rails and rowing seats. The rowlocks could be standard but the riggers they fitted on we would have to make ourselves. The towing eye at the bow we would likewise have to make. And finally I wanted *Yantu* to be bright orange above the waterline so that she would be easy to see, not that I was pessimistic or anything. Feeling quite pleased with myself I sent the fax.

The first week of February was my week off from consulting and I went to China to check how work on *Yantu* was progressing and train with Sun Haibin for the first time. Sun Haibin arrived on the train from Beijing with good news. Yu Zhen and a cameraman from CCTV would come down to visit at the end of the week. Mr Guo was well pleased at the prospect of having national television visit his factory. As the week progressed an increasing number of oversized

Completing the hull

red banners with big white Chinese characters was commissioned and hung up, some all the way from the third floor to the ground.

The Chinese love big banners. The banner reads: "Quality is the source and mainstay of Luyang's growth".

Sun Haibin and I shared the guest room at the factory, which was functional, but a far cry from the five-star hotels I stayed in when I was working. The floor was tiled, the walls were whitewashed, the en-suite bathroom had a hot shower as long as you shook the gas bottle and the beds were queen size, but without mattresses. We had breakfast with the workers, which mainly consisted of rice gruel, a watery substance which I really do not like, but Sun Haibin thought was top notch. Finding food we both liked for the race was going to be a challenge.

Sun Haibin had put a training programme together and every morning we got up at six and ran for an hour. The area around Hong Hai Bay was beautiful. The factory was on a bay full of fish farms and boats that looked like ancient bamboo rafts, though they were really made of PVC pipes sealed at each end. One morning we ran along the beach and came across a half-open makeshift tent. It had a very dead man lying inside it. We later found out he had died in a traffic accident and was placed there until he could be buried. I was a bit shaken, but Sun Haibin found it normal.

We continued our run along the beach and passed an impressive-looking building. This was a government-run sail-training centre, which some of China's best dinghy sailors and windsurfers use. I knew some of my old friends from the Etchells trip were training there.

Further along the beach we ran up the sand dunes and along the coast among a great number of family graves, which had been placed there because of the good *feng shui*. It was quite beautiful and tranquil, until an unfriendly dog came charging at us. We turned around and headed inland as fast as we could. We ran back through the village of Hong Hai, the centre of which is right out of a picture book of idyllic China. We passed an ancient temple of the Goddess of the Sea, Tin Hau, and re-entered the factory compound. We had been running for an hour. We did some push-ups and sit-ups and then went for breakfast. Work started at eight.

Given we only had one month until the launch, there was a lot left to do. The hull was not completely finished and we still needed to do the foam filling. Sun Haibin and I mucked in with the workers. I was impressed with his dedication and care in his work and was feeling increasingly good about him. He was also a very good team player.

The worker responsible for the towing eye at the bow was having a hard time. There were no power tools to cut the stainless steel so he had to make the eye and supporting plate by hand. This took a long time, but eventually the job was done. The finished result looked great and was certainly strong enough for its intended purpose.

We were making good progress, but I was concerned that work had not started on the oars and the riggers. Mr Guo told me not to worry. There was still plenty of time. I was not so sure, but he was the boss.

Yantu was sitting on a number of tyres on a cement floor. To fit the rudder we had to lift her off the ground as the rudder would protrude below the skeg, by not more than 20 cm as specified in the race rules. A number of workers were called and *Yantu* was lifted up, the necessary measurements were made, and she was put back down on the tyres. Sun Haibin and I then started fitting the steel braces that the rudder would fit in. We upped the dimension of the bolts one size to make the fittings extra strong. Losing the rudder would most likely mean we would have to retire from the race as we would not be able to keep *Yantu* on course. We worked very diligently to make sure the fittings were top notch and were quite pleased with the result. We then turned our attention to the next issue. Somehow we had to be able to control the rudder from our rowing stations. Keith suggested running a steering line from the rudder, along the outside of the boat, and into the cockpit through the scuppers. We could then work the rudder and cleat the steering lines off without leaving our seats or relying on help from each other. It was a simple solution, which meant it would be easier to repair if it malfunctioned, so we went about running the lines and fastening the cleats.

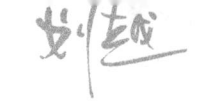

Yu Zhen and the cameraman arrived Thursday evening. Mr Guo considered them VIP visitors, as it was the first time that national television had been at his factory, and arranged for a luxury car to pick them up from the train station in Shanwei. I looked at the car. It was a right-hand drive! Cars in China are left-hand drive. A lot of luxury cars disappear from Hong Kong and are smuggled into China. Here was one of them, it seemed.

UPDATE

> In the past 19 years, China has gone from having no private car ownership to being the world's largest automotive market, both in terms of sales and manufacturing. In 2019 there were 21 million new cars registered in China versus 17 million in the United States. Western car manufacturers used to bring obsolete models and production lines to China, where they could continue to produce and sell models that were no longer desirable to Western consumers. Chinese consumers gladly bought these since they were advanced compared to what the Chinese domestic car industry could turn out. This is no longer true. Western car manufacturers now develop cars specifically for the Chinese consumer, who has become very demanding. China's own domestic car industry has also taken off, turning out quality cars that are starting to be exported. Western car manufacturers are still locked into compulsory joint ventures with Chinese partners. This "mutually beneficial" cooperation has been critical to the development of the domestic Chinese car industry, which would otherwise have been blown out of the water when obsolete Western models started to be produced and sold in China. In 2010 Geely acquired Volvo cars, signalling that Chinese car manufacturers have gone from

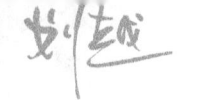

being plan economy basket cases to formidable companies with international ambitions and financial muscle.

After dinner with Yu Zhen and the cameraman we arranged to meet the next morning at six o'clock for them to film us training. They then headed back to their hotel and Sun Haibin and I returned to the factory.

The next morning we got up at 6AM and went for our run. The cameraman shot a lot of footage of us running up and down the beach in the rising sun. It felt like being in a B-movie with only the music missing. Having completed filming us training, Yu Zhen wanted to interview the Luyang staff who were working on the boat. *Yantu* was crawling with workers all hoping to be interviewed. Yu Zhen was very impressed by the fact that it had been put together without the use of a single nail. She asked Keith whether he thought we would make it. 'The boat is strong enough, but it remains to be seen whether they are,' replied Keith.

(Left) Reporter Ye Zhen from CCTV 5 making her reportage. Keith Mowser and me in the background. Yu Zhen and I ended up becoming friends. She later studied in England and in 2002 we visited Atlantic College together. (Right) The numbered segments outline advertising spaces we hoped to sell to companies in order to raise scholarships. Easier said than done.

Yu Zhen then asked me if I had finally settled on Sun Haibin as my rowing partner and I said yes. Sun Haibin may have been the only candidate, but we were starting to get on really well and we had the same kind of humour. Ask anyone who knows me and they will tell you that I have a very weird kind of humour, so this was a real plus.

Mr Guo wanted to put his logo on the hull when CCTV came to film and I had agreed that he could put on a small one on the unpainted hull, although this was not part of our deal. On the day of the filming

Completing the hull

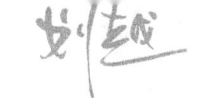

he had shown up in a new suit, but he did not get interviewed to any great extent. I could feel this was bothering him. CCTV left again the following day. Mr Guo insisted on paying for their hotels and presenting them with a *hong bao*, a red envelope with money in it, mainly given at Chinese New Year. Yu Zhen and the cameraman were embarrassed. I did not like it either as it flew in the face of the project philosophy of working on goodwill, which I told Mr Guo, but he insisted it was local custom and not to worry. The programme aired two days later, again five minutes on CCTV Sports News watched by about 50 million viewers. It was a good programme, when you ignore the embarrassing karaoke-like footage of Sun Haibin and I running across the beach. Luyang got a mention as well. However, Mr Guo was not pleased with his return on investment.

Shortly before leaving to go back to Beijing, Sun Haibin dropped a bombshell on me. He told me he might not be able to row to Hong Kong next month because there were problems with his travel documents.

'What do you mean?' I asked, breaking out in a cold sweat. 'Both the China Rowing Association and Zhang Jian have always said this would be no problem!'

'That is what I thought, but things are a bit more complicated. The China Rowing Association did not simply pass the project to Zhang Jian because of the clash with the All China Games. The State General Administration of Sports did not approve the project because they are worried we might die. That means the China Rowing Association cannot actively support the project. Unfortunately, they forgot to inform Zhang Jian about this when they passed the project to him.'

This was very bad news! I had thought government support was in the bag. In China, if you travel on official business, the process of getting a passport, being able to take leave of absence, applying for a visa and pretty much anything else connected with travelling is easier and faster. Sun Haibin would now have to apply for everything as an individual and have to battle with the bureaucracy and face time delays associated with this. And he only had one month to get it all done, which seemed impossible.

In order to apply for a passport, he would require a document from his university stating why he needed it, that he had a clean criminal record and was a good citizen. He would also need to have leaves of absence approved for the following periods during his final year of university:

- 8–11 March: Rowing the boat from Shanwei, China, to enter Hong Kong by sea.
- 28 April–6 May, 2–10 June, 30 June–8 July, 11 August–22 September: Training and preparation in Hong Kong.
- 22 September–7 October: Preparation in the Canary Islands.
- 7 October to about 1 December: Rowing the Atlantic, expected to take 50–60 days to the finish in Barbados.

Without the Administration of Sports' endorsement what were the chances that the university would approve his participation in the race? Things were suddenly looking grim.

The only thing I could do was to call Zhang Jian and impress on him the importance of what was at stake. Sun Haibin needed to be able to go to Hong Kong in one month's time, come hell or high water!

Before leaving, Sun Haibin and I agreed that he would find a rowing coach and a Concept II rowing machine in Beijing and learn how to row before we would meet again in March for the launch.

I watched him disappear down the dirt road on a three-wheeled motorbike taxi heading towards the train station in Shanwei and then walked up to the sail-training centre to see if I could find some of my old friends.

I knew that Shen Sheng, China's number-one Laser sailor, was interested in becoming my rowing partner, so I looked him up. I also knew he already had a passport because he had competed in several international competitions abroad. Could he be the solution? Shen Sheng and I talked all night.

Shen Sheng was slightly taller than me, and a stronger build. Before becoming a Laser sailor he had been a rower. This was clearly in his favour. Unfortunately, he was also tipped to win the Laser-class gold medal at the All China Games and, while this endorsed his abilities, dedication and potential as an athlete, it was a problem for his participation in the Atlantic rowing race.

The sports industry in China is very different from the West at the amateur level. In the West we can go to public sports clubs and try out any sport we fancy. When we find one we like we stick with it and if we really have potential we may turn professional one day. China does not have many public sports clubs and those that exist offer only a very limited range, mainly running, swimming, gymnastics and other sports that do not require expensive equipment. Rowing and sailing require expensive equipment so the average Chinese does not have the opportunity to have a go at it. Yet China competes in the Olym-

Completing the hull

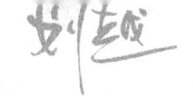

pics in these disciplines, so where do these athletes come from? The answer is that they are talent spotted at school, not for their proven ability or passion for the sport, but for their perceived potential. So, if you have the right physical build, you may be a normal student one day and then told you are going to be a sailor the next, though you may have never even seen a performance sailing boat before.

Once in the system they are stuck until they have outlived their careers. Athletes may therefore end up doing a sport they do not like and in an environment they do not appreciate. Rowing and sailing are not sports of choice, because they are physically demanding, outdoors and make you tanned. In the West we happily buy tanning lotions and toast on the beach, but in China whitening lotions are the rage because being pale signals you do not have to do physical labour. Being pale is associated with wealth as it once was in the West.

On the other hand, the Chinese and Western sports industries are similar. Once you turn professional (in China by decree and in the West by personal drive) you are expected to win medals for your country, club and investors, and you are bound by contract to that extent.

Of course contracts can be broken, but it seemed unlikely that the province that sponsored Shen Sheng would let him drop out of the China National Games without repercussions when he was expected to win a gold medal (which he subsequently did). He might personally be willing to forgo his cash prize, but it was unlikely his province would feel the same way and they would probably try to shut down my project. I could not afford to, nor did I want to, upset the Chinese government so in the end Sun Haibin was still the better candidate, even if it was touch and go whether he would get a passport. He was a student and not obligated to anyone to win medals.

Sun Haibin and I had been on the national news twice, surely this was some kind of endorsement, I reasoned. His passport simply just had to work out. I said goodbye to Shen Sheng and went back to Hong Kong.

I received an e-mail from Bob Wilson, President of the Hong Kong, China Rowing Association, who had read about my project in the December issue of the Royal Hong Kong Yacht Club magazine, *Ahoy!*. He would be pleased to help if he could, he said. We arranged to have lunch at the Yacht Club.

Bob is an elderly British gentleman with a long history of rowing feats, I learned over lunch. He had once rowed around Hong Kong Island, a distance of 25 nautical miles. He was keen to share his experience.

'Equipment is very important,' he said with great seriousness. *Pretty obvious,* I thought, not knowing where this was going.

'Particularly the seat,' he continued.

'Sure,' I said, not very interested.

'It is!' he insisted, looking at me with knowing eyes. He shifted forward on his seat and looked around as if to see whether anyone was listening. I shifted forward too. In a low serious voice Bob continued: 'Your ass. It will hurt beyond belief! And not only that — it will change shape! Your sit-bones will push through your muscles and through the holes in the seat. You must find a good seat!'

Somehow it seemed out of place that this softly-spoken elderly British gentleman was preoccupied about the future state of my ass, but I thanked him for the advice. I then told him about the pressing issue of Sun Haibin's travel documents and asked if he would mind writing to the China Rowing Association to officially request assistance. Bob promised to get on to the case and a few days later I received a copy of his letter.

Bob had pulled out all the stops and I could not have asked for better support. Whether it would help remained to be seen.

Martin Reynolds, the Royal Hong Kong Yacht Club rowing captain, had also read my article and he contacted me to ask if the Yacht Club could be of help once *Yantu* arrived in Hong Kong. A sponsorship agreement was struck. The Yacht Club would provide a stand for *Yantu* at Kellett Island, take her in and out of the water for practice sections, order equipment (I would pay for it) and install it for free in order to get *Yantu* ready for the race. The Club would also help with PR, fundraising, and allow me to write regular articles in *Ahoy!*. In return the Club would get a large logo on the hull of *Yantu* in a prominent place on both sides. This was serious help and a great deal for me!

The boatyard manager Duthie Lidgard, an experienced ocean racer and fourth-generation yacht designer and boatbuilder from New Zealand, became my key contact at the Yacht Club, so I was in capable hands. He borrowed Rob Hamill's book *The Naked Rower* and read it overnight. Being a fellow Kiwi, Duthie knew many of the people who had helped Rob in his 1997 entry and he contacted them to try and get some insights on equipment. He also contacted Rob himself, who came back with a positive response, despite us being competitors. Rob knew we were not a serious contender for beating his 41-day record set in the 1997 race, so there was no risk for him in helping out.

Completing the hull

Guy Nowell, another Yacht Club member and a professional photographer, contacted me to ask if I needed a photographer for the launch in China. It would be at no cost to me as long as I credited him when I used the pictures. Another great help!

My friends Francine Kwong and Tammy Wan from Atlantic College were at the early stages of preparing a fundraising dinner for United World College alumni in Hong Kong and were making good progress. Peter Davies, yet another Yacht Club member, said he had some time as he was out of a job and offered to help contacting local companies for fundraising. Maybe he could find a good job in the process, he reckoned. Kate Vernon, a friend of mine from Durham University had just arrived from England with her husband Matt, who was going to work as a vicar for the Church of England in Hong Kong. Kate had some spare time on her hands so she volunteered to help with fundraising, too.

In short, things in Hong Kong were progressing pretty well. In between weeks of work in Singapore, I made several trips to the Luyang factory to check on progress. Keith had finally managed to find the Lewmar hatches and the painting of *Yantu* was completed.

'Yantu' in the process of being spray painted. The red characters on the wall translate as "Challenge the Atlantic Ocean", "Challenge Nature" and "Challenge yourself".

The China launch

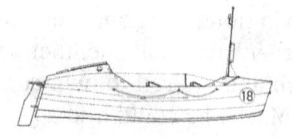

I arrived at the Luyang factory on Saturday 3rd March, five days before the launch. Sun Haibin arrived the following day and we settled into our on-site luxury accommodation. The passport problem was still not solved, but Zhang Jian was working hard at it. Hopefully when Sun Haibin's girlfriend arrived the day before the launch she would bring his passport along, which would allow him to row with me to Hong Kong. We turned our focus to getting the boat ready, while hoping for the best.

The oars had finally been completed, which was good news as it would have been hard to row *Yantu* at the launch without them! The bad news was that they were too thick. They had been built over a normal mould, designed for three layers of carbon fibre. When building our oars, the core of the mould had not been reduced and therefore, when five layers of fibre were applied, the end result was neither the sleeve nor the button fitted properly. To get a steady stroke the sleeve around the oar should fit snugly into the rowlock at a right angle, but now it could not rest properly. It was difficult to keep steady. Mr Guo insisted the oars were built to my specification. There was no way we could get any other oars in time, so we would have to use the oars for the launch and then get proper ones later.

The next task was to rig *Yantu*, i.e. to fasten the riggers, rowlocks, slides, seat and footrest. Chris Perry, Hong Kong's Olympic rowing coach, had promised to come to China to help with this, but unfortunately he had to cancel at short notice. This was not an immediate problem, as the riggers, seats and footrests were not yet ready so there was nothing to fasten just yet. Again it would not be much of a launch without seats and riggers, so the pressure was on.

Now the real reason why these items had not been produced ahead of time surfaced. They did not know how to make them. I did not know either. What to do? Frantically, I took out Rob Hamill's book, turned to the pictures and pointed at a rigger. 'Go and make something like that. Stainless steel,' I said to the metalworker. He took the book and walked off to his workshop. Next problem. We fixed the slides onto the two hardwood beams running the length of the boat, or we tried to. There were no bolts long enough to fit through the beams. Another metalworker was called to custom-make the bolts. A few hours later the bolts were ready and the slides fixed. Next problem. The rowing seats. A woodworker was called and it was explained to him what the seats should look like. Next problem. The distance between the slides was wider than standard and the normal wheel attachments to screw onto the bottom of the seat did not fit. Another metalworker was called. And yet another was working on making the footplates. In between checking on progress Sun Haibin and I screwed in the Lewmar hatches. The progress on the riggers was particularly slow because they were made of stainless steel and had to be handmade because of lack of power tools. During the occasional light moment I would feel proud that *Yantu* would probably be the only race entry with truly handmade riggers, but mainly I felt sick to the stomach that we would not be ready in time. The atmosphere was incredibly tense. We had to fix lights around the boat so everyone could work late into the night. For three days leading up to the launch we worked till 3AM and started again at 8AM.

As we were entering this crunch time the media started to arrive. CCTV from Beijing, Guangdong Province Television and Guangzhou City Television, the latter two due to the efforts of Sun Haibin. We finally had more press than we could shake a stick at, but no time to spend with them because the boat was nowhere near ready for the launch. We would almost forget that Sun Haibin still did not have his travel documents for Hong Kong.

I did an interview with CCTV showing them the route we planned to row down to Hong Kong. 'We will row along the coast. About here we will leave China and enter Hong Kong,' I explained enthusiastically tracking my finger over the chart for the benefit of the camera.

'Cut!' Yu Zhen shouted, laughing. 'You can't say that. Hong Kong is now part of China. You are not being politically correct!' We did a retake.

Back to working on the boat. The riggers were now ready, but where to put them? None of us had any idea. I tried to call Chris Perry. No reply. Panic! Who will know? I called Rob Hamill on his mobile phone.

'Hi Rob, my name is Christian Havrehed and I will be a competitor in this year's Atlantic Rowing Race. I got your number through Duthie at the Royal Hong Kong Yacht Club. I got a problem. I don't know the measurements for fitting the riggers. You have done this so I wonder if I could maybe have your measurements?'

'Hi Christian. No problem, but my boat is in a garage about two hours' drive from here. I will have to get back to you. You know, it is really not that difficult. Any rower will know how to,' he replied.

'Yes, of course. Thanks anyway' I said and hung up. I could not bring myself to tell him how desperate I was and that I was not a rower and did not have a clue. For the next hour I frantically called anyone I could think of. It was dark when I finally got hold of Chris Perry.

'Chris. Thank God! Look I'm desperate. I need to fix the riggers, but don't know how to. I am sitting in the boat. You have to talk me through it right now!'

For the next half hour Chris answered all my, to the most basic oarsman, stupid questions. I thanked him profusely and hung up. 'OK, we drill here,' I said to Sun Haibin and pointed to the side of the boat. Very early in the morning the day before the launch we were finally ready. The row stations were operational, but there was no doubt they would need to be re-done and some of the equipment replaced when we got to Hong Kong.

The day before the launch was spent buying equipment for the boat. The race rules required 150 litres of fresh water to be stored in jerry cans in the compartment under the stroke seats. This was for two purposes. It provided the weight needed for *Yantu* to be self-righting in case of capsize and it also served as emergency rations in case our desalination plant broke down out at sea. It should be sufficient to keep us alive until we were rescued, if it came to that. We went hunting around Hong Hai village looking for jerry cans and eventually found some the right size. We got them stored in place after first washing them out with boiling water to prevent the freshwater from getting contaminated by the plastic taste over time.

With half a day to spare, *Yantu* was now ready. Mr Guo had decided to build a ramp so that she could be launched in style in front of all the VIPs and television cameras. Instead of a standard slipway leading from the pier into the water, two metal tracks continued off the end of the pier straight into the air about three metres

The China launch

over the surface. The concept was that *Yantu* would be pushed out onto these tracks on a wheeled cradle. Once she was fully on the tracks, straps would be lowered around her. The straps were connected to a block suspended from a metal bar, which in turn was connected to a winch on land. *Yantu* would then be lifted clear of its cradle using the winch, the cradle would retract, and the boat would be lowered into the water in between the tracks. A lot of things could potentially go wrong. The cable might break or *Yantu* could slip out of the straps. In either case the result would likely result in significant damage. It was decided to try out the contraption just to confirm it did work. It did, but it was very stressful and Sun Haibin and I had to stand in the boat to fend off the metal pillars holding the tracks as *Yantu* was lowered and raised.

The launch ramp under construction. That is China in a nutshell. Let's just whip up a launch ramp in no time and lower the boat into the water with a winch. Looks much better than using a slip way. The banner on the left reads "We wish rowing boat 'Yantu' fair winds". And the one on the right reads "Company 'Luyang's' boats cross oceans to reach land".

The highlight of the day was the arrival of Sun Haibin's girlfriend, Cao Xinxin, from Beijing. She had brought his travel document!

On the night before the launch we were stowing provisions for the estimated three-day trip down to Hong Kong. Mr Guo came along to see how we were getting on and with a request. Since there would be so much press at the launch next day, could he paste his logo down the side of *Yantu*? I replied this was not part of our agreement and if *Yantu* was shown with his logo it would look like we already had all our sponsors, which was certainly not the case. But I could offer him a permanent space on the hull, if he would halve my bill. He was not willing to do this, so we did not reach an agreement. However, I did offer to put his logo on our shirts. After a very exhausting day Sun Haibin and I finally went to bed.

We woke up early on launch day. The weather was foul. It was raining and blowing a force 8 gale. The marine weather forecast was reporting a 'rough to very rough' sea state. I asked Keith for advice. He is an avid sailor and knows the coast well. We were standing looking out to sea. In the bay there were no waves because it was well protected, but Keith pointed to a coaster out to sea heading for the bay. 'See how she's pitching. She's coming in into the bay to shelter from the weather. You could maybe manage the row, but it's rough out there.' It seemed crazy to try and row in such conditions when we did not know the boat and all the equipment was untested. In addition, ocean rowing boats are dangerous close to the shore because they have a high freeboard and therefore drift significantly with the wind. With a gale blowing on shore we would almost certainly be smashed onto the rocky coast and *Yantu* would be lost. We decided to go ahead with the launch, but reluctantly cancelled the row to Hong Kong.

I had asked to follow local customs for the launch. At eight o'clock Mr Guo's elderly father came to collect us and we headed off to pay our respects to the sea goddess Tin Hau at the local temple we had passed on our morning runs. We offered a roasted pig to the goddess, donated money, and burned incense. The ceremony ended with Mr Guo's mother throwing two half-moon shaped wooden divination slips — one ended face up, the other face down; our boat had good luck! The whole experience was very emotional and Sun Haibin and I left with a lump in our throats.

The China launch

Feeling a lot better we walked back to the launch platform. *Yantu* was still in the courtyard and we arrived just in time to see staff preparing to stick large Luyang logos down the side of the boat. I had to forcefully prevent them from putting on the stickers. It was not a good situation.

The launch itself at eleven o'clock was fantastic, despite the rain. CCTV, Guangdong Province TV, Guangzhou City TV had now been joined by Shanwei City TV. In addition, there were journalists from Hong Kong and several local VIPs. Around 100 people had shown up, including immigration officials wearing black uniforms with silver buttons waiting to clear us out of China for the row to Hong Kong. They had brought a table with them and lined up their chops ready for action. Seeing how much effort they had put into the preparations, I felt bad telling them that due to the weather we would not be leaving by sea.

Firecrackers were set off and we had the good luck to have the mayor of Shanwei and the official responsible for the All China Games water sports unveil the boat, which was covered by a red silk cloth.

Yantu was rolled out onto its launch platform, the straps were put under her and ... she failed to lift up. The winch was broken. Embarrassment followed as the engineers went about fixing the problem while the VIPs got increasingly wet. Eventually the winch was abandoned and *Yantu* was raised and then lowered by hand. Due to the strong wind on shore, we had arranged for the press boat, a small dinghy with an outboard engine, to tow us out to sea a bit. The dinghy sped off at full throttle and *Yantu* smashed straight into a floating fish farm. Luckily, apart from a 10 cm scratch down the side, we suffered no damage. We let go of the tow and Sun Haibin and I could finally do what we had wanted to do for ages: row our boat.

Yantu performed exceptionally well in the water and we spent the next hours rowing up and down the coast and becoming more familiar with the boat, closely followed by a dinghy full of TV cameras. We could already make a few observations. We could just about row in wind force 5-6, but when it was gusting 8 we could not really do much! If this happened in the Atlantic it would be time to put out the sea anchor and climb into the cabin for cover.

Getting lowered into the water on launch day was a hairy experience. First the winch lowering us got stuck and then we were towed by the boat behind us and smashed into the fish farm seen in the background.

The China launch

Our first row in 'Yantu', the first ocean rowing boat built in China - and Asia.

Back on land again, CCTV had a request. They had not got enough footage of us rowing and since we were not rowing down to Hong Kong, would it be possible for us to go to Shanwei rowing club to practise some more? Given the mayor was present, this was a simple request to carry out. While the rest of the launch party went to a celebration banquet, we piled into a minibus and headed for Shanwei.

We arrived at Shanwei Rowing Club and realised the only boats for us to row were double racing sculls.

Yantu is 7.2 metres long, with a width of 1.8 metres and a freeboard of about 1.4 metres. She weighs about 600 kg, including the water ballast. You can move freely around her and stand on her side without capsizing her. You do not need to feather the oars on the return stroke, so you don't need much technique to row her.

A double racing scull is about the same length as *Yantu*, but only 35 cm wide with less than 10 cm freeboard. It weighs about 30 kg and is completely unforgiving. You cannot row it without feathering the oars and technical ability is a must. It capsizes straight away if you don't know what you are doing, particularly when it is blowing gale force 8.

Sun Haibin and I did not know how to row a racing scull. I had only rowed play sculls and Sun Haibin had started his rowing career earlier that day by rowing *Yantu*. We both knew that the scene about to unfold in front of CCTV's camera would not be pretty! We managed to get into the scull while the local rowing team were holding the boat, but we could not pull away from the jetty. We were too unstable. It was decided that one of the great Atlantic rowers better get out and be replaced by one of the club team. I watched Sun Haibin and the coach pull away. Pure terror on his face. They rowed for about half an hour by which time Sun Haibin was able to row with his arms and a little bit with his legs. Then I had a go with the coach. It went a bit better as I had had experience in the play scull, but I was not happy either.

Eventually, Sun Haibin and I got into the scull together again. With great difficulty and looking very wobbly we pulled away from the jetty. On land the cameras were rolling, capturing our every move. We were 100 percent focused on not capsizing. We tried to work up speed. People on land started cheering and running with us. *Great to have such a cheering crowd,* I thought! The yelling got louder and more urgent until finally we looked over our shoulders. We were metres away from ramming a barge. 'STOOOOP!' Sun Haibin yelled. We forced the oars backwards in the water and stopped just short of disaster. Now at a standstill the scull was even tippier and the gusting side wind was begging us to capsize. Slowly we managed to turn around and by pure luck we managed to get back to shore without capsizing. The whole thing had been a complete disaster. CCTV footage was only good for one thing. It would be a clear winner of the 'funniest videos' award: Look at these two idiots preparing to row the Atlantic!

By the time we got back to shore, CCTV had left. The club rowers told us that the cameraman had at one stage put down the camera and announced that we would never get across the Atlantic and they thought he had a point. I called Yu Zhen from my mobile and begged her not to show the footage. 'No one will ever sponsor us if you show that. The whole project will collapse,' I said. 'Besides, it does not matter that we cannot row a racing scull, we will be rowing *Yantu*, which is completely different.' Yu Zhen did not sound too convinced, but promised to see what she could do.

The China launch

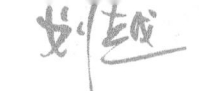

The disasterous row in a double racing scull as shown on National television. With no previous experience or technique we looked and felt like complete idiots. A terrible end to our launch day.

Sun Haibin took the whole thing remarkably well. 'I really learned a lot today. I did so badly. I have to practise a lot or I will never be able to do the row,' he said. It is impressive for anyone to make such a frank admission, but for a Chinese to so completely lose face and still be prepared to go back for more was truly something special. That day I gained a lot of respect for Sun Haibin. I realised he was as tough as nails and totally focused on completing the race.

That evening we took the Luyang people out for dinner to celebrate the launch and thank them for their work. They had done an excellent job of completing the hull, but no amount of beer and karaoke could disguise the tense atmosphere between Mr Guo and me.

The next day Sun Haibin and I went by road to Hong Kong. *Yantu* was put on the back of a lorry and would arrive at the Yacht Club two days later.

Our five minutes on CCTV showed pictures from the launch, and us rowing the racing scull. At closer scrutiny it consisted of the same one-second clip repeated several times. We looked almost competent and the news reporter said we had 'managed to row in a basic way'. I thanked my lucky stars for Yu Zhen. It could have been a lot worse.

My March e-mail update to the Yantu Project supporters read:

It is a good feeling to have the boat in Hong Kong and be able to go down and have a look at her whenever I want to. Given that we now have a ship shop right next to the boat, whereas the nearest one in China was a few 100 km away, things can now be done much faster.

We are now turning a new leaf in the Yantu project — row & raise sponsorship.

Training and fund raising

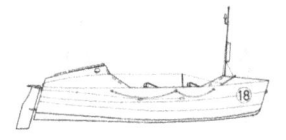

Sun haibin had only been outside Mainland China once before when he went to Macau for an iron-man competition, so Hong Kong was a completely new experience for him. After dropping off our bags at my flat and giving Sun Haibin a key so he could come and go as he pleased, I wanted to show him the Yacht Club, which would be our training base from now on. It was dark by the time we arrived and we decided to head for the bar.

The music was blaring and the bar was heaving, mainly with Westerners. Girls with blue wigs, sporting propped-up breasts, low cleavages, miniskirts and fishnet stockings were serving beer much to the enjoyment of everyone. We had walked right into a fundraiser for the San Fernando Race, an annual yacht race to the Philippines, which raises money for an orphanage there. I looked at Sun Haibin. I knew he had only had limited interaction with Westerners and maybe this was not the best way of introducing him to the Western way of life. It is difficult to say what went through his head, but he looked pretty culture-shocked. Duthie, the boatyard manager, spotted us and, beer in hand, fought his way through the crowd to greet us. We went outside to talk as it was impossible to hear anything inside. Duthie was very welcoming and Sun Haibin soon started to feel relaxed, though he couldn't understand a word Duthie was saying. Duthie explained that everything was ready for receiving *Yantu*. They had converted a metal delivery stand for a Beneteau yacht into a cradle and put some wheels on it. This was exactly what we wanted to hear.

Yantu arrived the next day and was put on the cradle. She sat about 1.2 metres off the ground. The cradle was all metal and very heavy, but it served its purpose fine. Unfortunately, *Yantu* had not been packed very well onto the lorry by the Luyang workers and had damage to her skeg. Nothing serious, but it prevented us from rowing her. Instead we went to the gym and rowed on the Concept II machines.

Later in the evening, Sun Haibin left to catch his train 3,000 km back to Beijing. He had volunteered to take the train instead of flying, as it was significantly cheaper. I appreciated this as I was paying for his tickets and the whole project was getting increasingly expensive. However, it meant that instead of a four-hour flight, Sun Haibin was preparing for a 27-hour train ride.

Before leaving, we agreed that Sun Haibin would go to a gym in one of the big hotels in Beijing to get some more rowing training on Concept II machines as he was still lacking upper-body strength.

While he was back in Beijing studying I went off for my three weeks of work. We met up in Hong Kong again for training from 7th to 15th April. I had arranged with the Yacht Club that we would do a Hong Kong launch of *Yantu* to try and attract the local media. And, since '8' is a lucky number, Sunday 8 April sounded like an auspicious date for the Hong Kong launch. But it rained! Nevertheless, due to the efforts of the Yacht Club's Sponsorship & Communications Manager, Ellen Wong, a lot of media showed up from both the English and Chinese press. The concept of having two different nationalities seemed to work!

The launch of Yantu in Hong Kong received solid interest from local media, both Chinese and English.

Training and fund raising

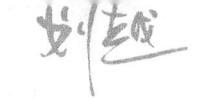

Yantu was hoisted into the water by crane and we got in. The press got into two Yacht Club motorboats. We headed out into the harbour. This sounds simple enough, but Hong Kong's Victoria Harbour is one of the busiest in the world. Tugs, barges, hydrofoils, ferries, coasters and fishing boats jockey for position in the narrow waters between Hong Kong Island and Kowloon. It is a place where size matters. If you are big you have right of way, if not you better get out of the way. We fell into the minute category, so we were on the lookout. We rowed at great speed straight out into the harbour, turned around and came back while the press boats followed us. Not too bad, we thought after the return to dry land.

Some of the press stayed around to do follow-up interviews and this was when I first met Victoria Button from the *South China Morning Post*. She did a very thorough interview covering the philosophy of the project and then she asked: 'Why don't you feather the oars when you row?' I was delighted at the question, because it showed she understood rowing. I explained that we had so much freeboard that we did not need to and also, by not feathering the oars, we were putting less stress on our wrists and were less likely to develop blisters. This was of key importance as our bodies would be under a lot of stress during the race, and everything which would prevent our bodies from breaking was important.

That evening we had significant coverage on the local TV news and the next morning a number of newspapers had features on us. We also had our first radio interview. The coverage had been both in English and Chinese, which was great news. We now had something to show potential sponsors!

I had to take care of some administration work related to the race and since Sun Haibin did not read or speak English and was therefore unable to help he decided to go to the Yacht Club to row on the Concept II rowing machines while I stayed at home. A few hours later he came back.

'The rowing machines don't work!' he proclaimed.

'What do you mean?' I asked.

'I rowed for two hours, but I never got any exercise. There is no resistance,' he explained.

I had been using the machines quite a lot so this did not make sense to me. The next day we went together to the Yacht Club to see what the problem was. I had asked my rowing friend Rob Stoneley to meet us there, just in case. Sun Haibin and I got on the machines and started rowing. The Concept II machines use a centrifugal wheel to provide resistance. The harder you pull on the handlebars the faster the wheel turns and the more resistance it creates. Sun Haibin was not pulling on the handlebars! No wonder he was not getting any exercise. His upper body was completely stationary while his lower body was going backwards and forwards like a piston.

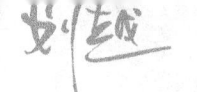

Jesus! I thought. *How can anyone have so little intuition? How are we ever going to get across the Atlantic?* I felt sick in my stomach and I had to ask for help. 'Rob, I can't handle this. Please teach him how to row.' I left the gym to get some fresh air. Some time later I walked back in. Rob was coaching Sun Haibin and he was now pulling the handlebars. I got on the machine next to him and fell into his pace. We continued for about an hour, by which time my gloomy mood had started to lift.

Some time later I got a call from Chris Perry, the Hong Kong Olympic rowing coach. His opening comment was:

> 'I saw you guys on TV the other day and thought I'd better come and give you a hand. I can't bear to think you will be flying Hong Kong's colours in an international event with that rowing technique.'

One early morning we then went out rowing with Chris as a passenger. He was standing in the well at the stern leaning against the hatch into the sleeping cabin and watching us carefully. His conclusion was that my technique was worse than Sun Haibin's, and neither of us showed much promise of becoming proficient before the race. He switched into damage-control mode and helped us correct some of our major technique shortcomings.

Chris' evaluation was not the best news, but I reasoned that he was a racing scull coach, so his opinion was not necessarily that relevant. He then went over the rigging and told us to get new oars, new seats, new sliding rails, new shoes, and to make sure the rowlocks were perfectly vertical in the riggers. This was valuable advice. Before leaving Chris volunteered to help us purchase the oars from Croker in Australia, who he knew had also made the oars for Rob Hamill.

Sun Haibin and I decided to venture further out to sea and row halfway around Hong Kong Island to the Yacht Club's clubhouse on Middle Island. This would be a long row in the busy harbour and for the first time we tried to row as we would in the race, one person at a

Training and fund raising

time. Sun Haibin took the first two-hour shift and I went into the cabin to rest. Some time later Sun Haibin yelled to me to come outside. From the tone of his voice I could hear that trouble was brewing and as I emerged through the hatch, the problem immediately became apparent. We had drifted between the pillars supporting a causeway, and were in imminent danger of being thrown against them by the wash from the ships further out. We managed to get *Yantu* back out into safer waters without damaging her and rowed together for a while.

We had stayed close to shore to keep out of the traffic, but now we headed further into the harbour in order not to hit land again. This meant we were in the middle of the traffic lane, but the ships never gave us a hard time. One fisherman sailed up to us. 'You are the guys rowing the Atlantic, aren't you?' he yelled across to us in broken Mandarin. 'I saw you on TV.'

'Yup!' we said, feeling proud. He then threw us two bottles of soft drink, waved and sailed away. What a nice gesture! We rowed on feeling great.

The row to Middle Island took seven hours. We left *Yantu* there overnight and rowed her back the following day. The wind had picked up and when we arrived off the village of Shek O, about halfway on our return trip, there was a two-metre sea. Sun Haibin had never seen waves this big before and was slightly nervous. I explained to him that as long as the waves were not white on the top and breaking, it did not matter how tall they were, they would always pass under us. That made him feel better, but they still made him seasick. The wind turned on the nose and picked up to a force 5-6. We rowed together against the wind for the next two hours. From taking a fix on land, it was clear we were not making any headway, but we were not getting blown out to sea, either. It was pretty hard and demoralising. We could only hope that such weather would be the exception across the Atlantic. Sun Haibin developed a nosebleed due to overexertion. He was obviously rowing to his full capacity. This got me worried, because conditions were not that bad. However, he kept rowing without complaining. *Tough as nails,* I thought. Eventually the wind died down and we could row back to the Yacht Club. We did a few more rows that week and then it was again time for Sun Haibin to take his train back to Beijing and for me to go back to work.

I spent three weeks in Croatia starting up a consulting project and on the way back I flew via Frankfurt. I wanted to stop in at Empacher, one of the world's leading manufacturers of racing sculls, thinking they should be able to provide me with all the gear I needed for the rigging. I spent a morning there buying shoes, sliding rails, rowlocks and seats. Because of the extra width of the slides on *Yantu*, they had to widen the brace with the wheels under the seats. I got some spare wheels in case we would wear them out. Despite the fact that we would be using the wheels in salt water, we opted for wheels with ball bearings, the reason being there would be little chance of them seizing up due to the constant use. It turned out to be a good choice. Feeling happy I flew back to Hong Kong.

The Yacht Club had ordered and installed a lot of kit on *Yantu*. She looked transformed. The solar panels were on and the electrics installed. Vents had been put onto each side of the cabin, as it would be hot rowing the Atlantic. The custom-made Weaver hatches for the two compartments under the rowing seats and for the storage compartment on either side of the well had also arrived and been installed. A marine battery had been put into the compartment on the left side of the well. The water-maker, which was still to come, would go on the right side. She increasingly looked like an ocean rowing boat. Duthie had ordered the electrics and hatches from the same suppliers Rob Hamill had used in 1997, so we felt confident it could go the distance. Things were progressing really well and I was looking forward to showing it to Sun Haibin. We had arranged he would come down for training from 29th April to 5th May.

The night before he was due to arrive, I received an e-mail message from Cao Xinxin, Sun Haibin's girlfriend. I had never had a message from her before and appreciated her getting in touch. I clicked on the e-mail, which was short and written in Chinese: *Sun Haibin is in hospital. He is not coming to Hong Kong.*

I felt like I had been hit on the head with a sledgehammer. I immediately called Sun Haibin on his mobile phone, which was switched off. I continued calling all evening, but it remained off. What was going on? He must be seriously ill, or he would not be in hospital! What if he wasn't in hospital at all and was simply using this as an excuse to cop out? It would not be unthinkable. The last training session had been tough. He had had a nosebleed and been seasick. Maybe he had decided rowing the Atlantic was not for him and he was using an indirect way of telling me this, as the Chinese commonly prefer, instead of direct confrontation. Eventually, I went to bed, but I could not sleep the whole night. I tried calling again the next morning and this time he answered.

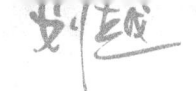

'Sun Haibin, what is going on?' I asked.

'I am in hospital. I have been here for three days and the doctor wants me to stay for another two,' he replied, sounding decisively ill, which, things considered, made me feel a lot better!

'What happened?'

'After I got back from Hong Kong, I knew I really needed to train my upper body so I went to the gym every day. I did not allow myself time to recover and wore myself out. I started to develop a cold, it turned into a flu and I ended up with a high fever, so the doctor put me in hospital,' he explained.

'Do you still want to do the race?' I asked.

'We have come this far and there is no way I am not going to do the race,' he replied with a firm voice. It made me feel good.

We started discussing what had happened to him and the important message to learn from his misfortune. We would have to listen more intently to what our bodies were telling us. The consequences of being macho and pushing on when the body said stop could be dire in the middle of the Atlantic. But we were both competitive and would both probably push ourselves too hard. We therefore decided that we would be responsible for each other's health and if one thought the other looked tired, he could ask him to rest. The other person would then have to do so. We both agreed that it would be better to slow down and rest than to push ahead.

'Anyone can sign up to row across the Atlantic, but it is getting all the way across that really counts,' Sun Haibin pointed out.

The cornerstone in the Yantu Project was to prove that it does not matter how different the cultures and how seemingly impossible the task, as long as both sides are willing to cooperate and trust each other, anything is possible. If we did not get across, sceptics could point to our project and say that working cross-culturally does not work. We did not want that! At the same time we both agreed we would not compromise safety for the sake of success. We would be rowing the Atlantic to live, not to die.

We decided that Sun Haibin should take it easy for a while once he got out of hospital and since we could not train together we might as well put more effort into finding sponsors. Sun Haibin would write to the people and companies mentioned in my friend Rupert's 'Hurun

China Rich List' and I would try and get Hong Kong going. I hung up. It had been a good and honest talk.

I decided to write a mail shot to all the members of the Danish Business Association, and then I called Peter Davies to see how he was getting on with the fundraising. Things were tough, he said. We borrowed the Yacht Club's sponsor list and together we went through it to identify potential candidates. Had we been doing this six months earlier we would have targeted the dotcom companies. In the meantime the bubble had burst and we were now striking anything that sounded like a dotcom from our list. Eventually we ended up with 115 candidates for whom we put together a detailed sponsorship proposal. The production of the letters took forever and after having licked 115 stamps and envelopes we were not feeling too good, but out the door they went. We agreed Peter would follow up in a few days with phone calls.

Kate Vernon came up with a good idea. 'Why don't you contact Rotary and ask to do lunch talks? You will have an affluent captive audience and Rotary and United World College are not too far from each other in philosophy,' she pointed out. I asked her if she could set it up and she went to work.

I called Francine Kwong and Tammy Wan to see how the UWC fundraising dinner was coming on. 'Very well,' the reply came back. We agreed to hold the dinner on Saturday 2nd June at the Yacht Club.

I arranged with Ellen Wong at the Yacht Club that we would do a media happening in June when Sun Haibin was back in town. We would row around Hong Kong Island. This had not been done since 1992, mainly because vessels without motors are not allowed in the western part of the harbour where the traffic is particularly intense. We wondered if the Marine Police would allow ocean rowing boats in the western harbour, but decided it was probably best not to ask. Instead we would time our crossing in such a way that we would avoid rush hour. Things were looking good!

The 2nd of May was my birthday and a small parcel arrived from my mother, who by now knew about the project. It contained a tin opener and a small note saying: *Happy birthday — just so the same thing does not happen to you.* It also included a newspaper article about Jim Shekhdar, who had just finished rowing solo across the Pacific in 274 days. Jim had forgotten to bring a tin opener, the article said. I thought that was an excellent birthday present. My mom was

71 years old and very much on the ball. The present signalled that she was coming to terms with me doing the race.

Peter was having a hard time with his follow-up phone calls. No one seemed to have received the letter. Probably the secretaries were throwing them out before they reached their bosses, so we decided to change the calling technique. Instead of calling to ask a given CEO if he had received the letter, Peter took a more direct approach.

'I am calling to discuss the letter I sent you a few days ago regarding sponsoring Hong Kong's entry into the Atlantic Rowing race. What do you think?' Peter would start out. Without fail the CEO had not seen the letter, so Peter would offer to fax it over straight away and then call back immediately. This technique worked better and it became apparent that we should have done this from the start instead of sending a mail shot. Without ever sending a letter we would call new prospects and ask them if they had received it. We would then fax it to them. In this way he managed to speak to quite a number of CEOs, so at least more and more of them became aware of the project, but money was not pouring in the door. The whole project seemed too difficult to sell.

'If you were fundraising to build schools in China, I am sure we could raise money. But you want to send Mainland Chinese students abroad to give them international education at the cost of US$46,500 for two years. You can build a lot of schools in China for that amount of money,' Peter commented.

In a way he was right, but hopefully the students we sent to Atlantic College would eventually become successful leaders and be able to put back into society much more than the cost of their education. However, selling this long-term concept was difficult. Putting myself in the shoes of a CEO, what would I rather write in my annual report? My company sponsored one student's pre-university education for two years at a privileged international school or my company sponsored the healthcare, food, education and clothing for 70 children in China for two years? I guess each NGO in a soul-searching moment may ask itself whether it really is the most worthy cause to sponsor. I don't think there is a definite right or wrong. Different NGOs have different purposes. Atlantic College aims to promote international understanding and that is a worthy enough cause to me. We would have to uncover CEOs who thought likewise. We pressed on with renewed effort.

It was time for me to get back to work. I jetted off to Indonesia and then Taiwan, where I would be working for the rest of May. I returned to Hong Kong each weekend to check on progress. *Yantu* was becoming increasingly race-ready, but we were not raising any sponsorship money. I really wanted to send a student to Atlantic College in September before the race to show beyond a doubt that we were serious about raising scholarships and that it was not a gimmick to line our own pockets, as some people seemed to think.

Sun Haibin came to Hong Kong from 2nd to 10th June for training. He had had equally little luck raising sponsorship in China, it turned out. We felt depressed. What were we doing wrong? But at least we had the UWC fundraising dinner that evening to look forward to.

We changed and headed down to the Yacht Club. Fifty graduates and a few potential sponsors showed up for the dinner and we had a great evening, which included a wine auction. Everyone seemed to enjoy themselves and some old classmates met up for the first time since graduating! Needless to say a few bottles of beer were drunk along with a lot of reminiscing. They all thought Sun Haibin and I were crazy, but that did not stop them from supporting our cause and we raised US$27,763 towards scholarships.

On the same evening, in response to my mail shot, we also received our first donation towards race costs from a Danish company called Delfi Tech Manufacturing Ltd. Because we were no longer bound by the UWC International Office's demand that all funds raised must go to scholarships I happily accepted the donation.

That evening was the watershed in our fundraising campaign. Afterwards, when we contacted sponsors we could tell them that we had already raised close to US$30,000 in scholarships, which gave us credibility. It also reassured the prospective sponsor that he would not be a laughing stock for sponsoring this seemingly crazy project as others had already done so.

A few days later, we saw this working in reality as we gave our first Rotary lunch talk and were immediately promised sponsorship assistance. We all kept up the hard work and slowly the money started coming in and sponsor logos began to appear on the side of *Yantu*.

A few days after the fundraising dinner we went for another row in the harbour, this time to Clear Water Bay Marina, nine nautical miles away. Since the round trip would only be 18 miles and take about six hours to complete, we decided to row together the whole time. It was tough. We had got new oars from Croker, which were a big improvement on the old ones, so that counted in our favour. However, the sun was burning in a clear sky and the temperature was around 30 degrees Celsius and it was punishing to row. We each drank one litre of water every hour and ate plenty of fruit. However, we ended up with nice tans after we stopped feeling crisp.

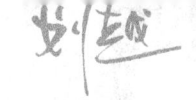

We took stock of the experience. The race across the Atlantic would take place in similar heat and we realised that training in Hong Kong was actually an advantage. Most teams were from Britain and they would surely suffer when they had to row in the heat. It also became apparent that we had to be careful to drink enough liquid and cover ourselves from the sun. There were no places to cool down on board if one of us developed heat stroke. The consequences of that in the mid-Atlantic were likely to be fatal.

A few days later we set out on our trip around Hong Kong Island. At 25 nautical miles this would be our longest row to date and it would include our first night row and cooking on board. We set out at 3:45PM in pouring rain and rowed west. This took us into the busiest part of the harbour, but again the other boats looked out for us. Nevertheless, we were thrown out of our seats once due to a big bow wave from a passing ship which turned *Yantu* on its side.

The weather continued to deteriorate and, as it got dark, thunder and lightning set in. The lightning was pretty handy because we had almost no visibility due to the rain. After two hours I dropped out to rest and Sun Haibin rowed for an hour. Then I rowed for an hour and we continued this pattern for the rest of the trip.

At 8PM we cooked our first meal on *Yantu* — pasta and mackerel in tomato sauce — and ate it in the pouring rain. Duthie had said to us before we left the Yacht Club: 'It's about time you start getting used to being uncomfortable — I'm off to a barbecue.' As we were sitting eating our pasta in the rain we could not help feeling a bit envious of him at the barbecue. Sun Haibin seemed to find the rain more difficult to endure than I and he was worried about catching a cold, something that seems to be a favourite Chinese pastime.

Eventually the rain stopped and for the last two hours we had perfect conditions. We arrived back at the Yacht Club at 1:50AM Wednesday morning, 10 hours and five minutes after we set out. The trip had not really been that tough and our hands were in pretty good shape. The rain had kept us cool and we had only drunk two litres of water each during the 10 hours, compared to one per hour during the day. The heavy tropical rain kept journalists from showing up at our press conference, but Victoria Button from the *South China Morning Post* interviewed us over the phone and wrote a nice piece about us the next day:

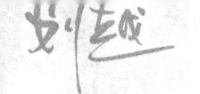

> *The 25-nautical-mile journey took them just over 10 hours — a drop in the ocean compared to the 50 days or more it will take them to row the Atlantic.*
>
> *'The rain keeps us cool,' said Havrehed after the round-the-island journey.*
>
> *When visibility fell to zero during a rainstorm, they navigated by using charts and a compass. Luckily their biggest enemy, the wind, was low. 'It went remarkably well. We had very good rowing conditions despite the rain, or maybe because of it,' Havrehed said.*

Yantu was now basically ready for the race. We were missing the safety equipment (life-raft and first aid kit), but Hans-Henrik Madsen from Danish company VIKING Life-Saving Equipment Ltd. was keen to help out, so we were not too worried. Apart from the safety equipment we only needed to install the water-maker and make the cabin comfortable, so readying *Yantu* was on schedule.

The week of training in Hong Kong taught us that what we drank and ate made a big difference to our performance. We decided to look into this.

Before leaving on his train back to Beijing, Sun Haibin informed me that he might not be able to come to Hong Kong for training in July because of his travel documents. I was completely taken aback as I had thought that was an issue of the past.

It turned out that Sun Haibin had been coming to Hong Kong on a business travel permit and not a passport. Back in March there had been no time to apply for a passport and the only possibility had been to get him a business travel permit. This had happened in a roundabout way. Sun Haibin was still a student and not a businessman, but a company agreed to 'employ' him, wrote and chopped a letter applying for a business travel permit allowing him to come to Hong Kong on company business. However, getting this letter was only half of the required paperwork. He also needed his *hu kou*, residence permit, to apply for the business pass. In normal circumstances this would be held by the employer, but in Sun Haibin's situation it was with Beijing Sports University. Sun Haibin explained to the university that there was no time to get a passport and therefore a business travel permit was the best way forward. At that point in time the university was still under the impression that the project was endorsed by the State General Administration of Sports and co-operated. Sun Haibin then went to the Beijing Police Administration and obtained a business travel permit, the inconsistency in his application papers going unnoticed. The permit only had three-month validity, so before coming to Hong Kong in July he would have to renew it and there was no guarantee he would be as lucky a second time around. As for the passport, which he would definitely need for participating in the race, this was not looking too good, either.

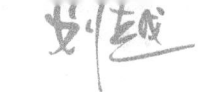

This was major bad news, but we both felt that if we had come this far, somehow it would work out. Sun Haibin got on the train and went back to Beijing.

Sun Haibin was scheduled to come to Hong Kong for training from 1st to 8th July, but called me with a mixture of good and bad news.

The good news was that he had finally managed to obtain a personal passport after endless amounts of trouble. He had needed the university's approval to apply, but they had by now become aware that the project was not endorsed by the Administration of Sports and therefore were reluctant to help him. He had shown the university all the press coverage we had received and told them everyone thought he was participating and it would not look good if he now did not go. After much pleading and writing a letter, complete with thumbprints, that if he died rowing across the Atlantic no one was to blame but himself, the university finally relented and supported his application. The process of getting the passport had then been quick and straightforward, but he could not use it to travel to Hong Kong. Believe it or not, Hong Kong, at the time, was one of the most difficult places for a Mainland Chinese to get a visa to!

The bad news was that the renewal of the business travel permit had failed. Around the time he was applying for it, a number of Mainland businessmen had been caught smuggling drugs back into the Mainland from Hong Kong. To tighten up on who travelled to Hong Kong, the authorities issued a regulation that the official who issued the business travel permit would also be implicated if the holder was subsequently caught smuggling drugs. Applications were therefore scrutinised more carefully and Sun Haibin's inconsistency was spotted. When he then went to try his luck he was not only turned down, but also questioned for two hours about his supposed drug smuggling activities by a hostile official! This was pretty serious as drug smuggling is a capital offence, and things did not look good for a while. In the end, he had produced the press cuttings to back up his story and the official could see that, yes, he was in fact preparing to row across the Atlantic! Sun Haibin was let off with a warning but had of course not been issued a new permit. To add insult to injury he had been stopped by a policeman on the way home and asked to show his ID card. Because of the rowing training Sun Haibin was deeply tanned and was therefore taken for a migrant worker. Going back on the underground the other passengers had moved away from him for the same reason. Even a little beggar boy had called him a 'bad egg' when he had gestured to him to come and sit next to him! No wonder tanning lotions are not the rage in China!

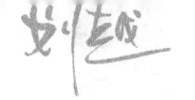

Sun Haibin was now in the bizarre situation that he could use his passport and apply for visas to go anywhere in the world, but could not come to Hong Kong. We took stock of the situation. Since we met in January we had only trained together for four weeks because we had missed May and now we would also miss July. Not a great track record for preparing to row the Atlantic. He simply had to come to Hong Kong for all of August, by which time I would have stopped working and would be focused full time on getting ready for the race.

For now Sun Haibin would have to stay in Beijing and visit embassies to put in his visa applications. The Challenge Business had been very helpful with providing the invitation letters that he was required to submit with his visa applications at the British and Spanish embassies and the Barbadian High Commission. The Hong Kong, China Rowing Association had issued a letter stating that I would be responsible for Sun Haibin's overseas expenses and make sure that he returned to China. Armed with all these documents Sun Haibin went to work and I concentrated on finding a way to get him to Hong Kong in August.

A friend of mine living in Hong Kong, Leo Austin, had a Mainland Chinese wife and, since love conquers everything, he had discovered a loophole. If Sun Haibin came to Hong Kong on a charter holiday he would be issued a special leisure travel pass as long as he arrived and left with the tour group. This sounded like a good idea and I told Sun Haibin to book himself on a charter holiday to Hong Kong. He found one that would allow him to come to Hong Kong from 1st to 12th August. This was not ideal, as we had counted on him being in Hong Kong until we flew out on 22nd September, but it was better than him not coming at all so I told him to go ahead and buy it. He did not have the cash so I had to ask a friend in Beijing to lend him money and eventually he got the ticket.

Training and fund raising

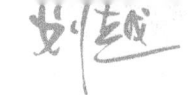

For a long time Sun Haibin and I had discussed whether *Yantu* should have satellite communication. The winners of the last race did not have any outside communication equipment, believing it would distract them from the task of rowing. Subsequently, they regretted this. It became too lonely and the atmosphere on board too intense.

We both favoured the idea because we reckoned we would sooner or later want to speak to someone in our mother tongue. It would be tiring for Sun Haibin to have to listen to my Chinese all the time and for me to speak Chinese all the time. In addition, satellite communication equipment would enable us to receive updates about the other competitors along the way and talk to journalists. It would also be an extra safety measure.

I was therefore more than pleased to find out that the satellite phone company Iridium was back in business. They were represented in Hong Kong by Stratos through STM Limited, whom I went to visit in order to introduce the Yantu Project. That resulted in a visit to the boat. One thing lead to another and Stratos and STM Limited ended up lending us a Motorola 9505 handset, a mast antenna and sponsoring 600 minutes of airtime. Great!

Hans-Henrik Madsen from VIKING came through and offered to sponsor our life-jackets, survival suits, life-raft, flares and medical kit. That was great news not only from a cost point of view, but as a Dane I felt an extra sense of security knowing that our safety backup was made by a reputable Danish company. An additional advantage was that VIKING could deliver our flares to the race start in Tenerife — they would otherwise be expensive and difficult to transport as they are classified as dangerous goods.

UPDATE

After the race I worked for VIKING Life-Saving Equipment as their China Managing Director for four years, during which time VIKING became the number one provider of maritime safety equipment to the Chinese shipbuilding industry and China became the world largest shipbuilding nation. We also established an extensive service network along the coast of China, enabling international ships calling China with VIKING safety equipment on board to have their products serviced to the required international standard. Who would have thought giving up my job as a consultant to row the Atlantic would end up landing me a great job as a Managing Director?

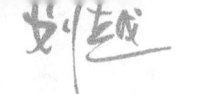

It was time to tackle the issue of provisions. For a long time I had had nightmare visions of spending August reading up on nutrition, about which I knew very little, going to the local supermarket to buy truckloads of chocolate bars and biscuits, and packing it all into day packs. I asked the Challenge Business for help and they put me in contact with Maggie Page, a British nutritionist specialising in extreme sports events. She had not only provided food for one of the Whitbread entries, but also for one of the teams in the 1997 rowing race. She told me that if we planned to row in two-hour shifts 24 hours a day, seven days per week we would each need a daily intake of 6,500 calories. (In normal life a person eats that many calories over two or three days.) That would equate to about three kilos of food per day for both of us. The race rules required we brought 90 days of food and Maggie proposed to prepare 60 days' worth of food at full rations and 30 days at half rations. The total weight of the food would be about 225 kg, despite it being freeze-dried. That was a frightening statistic. We were clearly training with too light a boat and would have to get more jerry cans and fill them with water so that we could train at race weight. I thanked Maggie for her information and asked her to send us some samples, which we would then try during training in August. As I was hanging up, I remembered to ask her to remove all cheese from the food because Chinese do not take well to dairy products.

Between sponsor presentations and administration, I did get some rowing done during July. I rowed out to Junk Bay to watch the match racing that marks the end of the sailing season in Hong Kong. The Yacht Club had rented a very large junk for its members and about 200 of them were watching the match racing while eating and drinking. I moored alongside the junk and got some food. A number of kids asked if they could come on board and for the rest of the afternoon I let them row *Yantu*. It was great to see their excitement. I dropped off the kids at the junk and a woman I did not know asked me if she could row back with me. Guy Nowell, who had previously photographed the China launch, also appeared and asked to come along. The three of us set off. I was standing in the well looking at them row. They were both quite good. The woman was called Lindie

Training and fund raising

Rudover and she was in Hong Kong for a short time, on a work assignment. I changed with Lindie and she stood in the well. Immediately she started correcting my technique. It turned out she was also a rowing coach. 'Don't use your back, use your legs,' she ordered and Guy soon joined her in correcting my technique. I tried my best to comply and after about an hour they stopped. Maybe the job seemed too difficult.

Back at the Yacht Club we went up into the bar. It was full with the people from the junk, which had by now returned. A very drunk bearded man came over and started laying into me.

'You will never succeed in rowing across the Atlantic. You can't row ...' he started. I looked at him and he continued: 'I know because I come from a family with a long rowing history. My father rowed and I row. Until you can row a racing scull quickly round a mark it does not count. I have rowing pedigree. You can't row. You will die rowing across the Atlantic!'

I looked at him and thought he did look like a Pedigree, though of the tinned dog food variety. Never before and — thankfully — never again did I get such a negative talking-to. Of course he had a point about my technique, but an ocean-rowing boat is a lot more forgiving than a racing scull, so I was not that worried about technique. I was convinced that if anything would let us down, it would be our mental state and not our rowing skills. I told him he was entitled to his opinion and after a while he staggered away.

Lindie was really excited about the Atlantic race and was feeling a bit envious that she was not competing herself. She knew Sun Haibin was not in town and therefore offered to come rowing with me any time. A few days later we set out to row around Hong Kong Island for the second time. No thunder or lightning this time, but we did have some excitement — we got stopped by the Marine Police who came charging towards us in a formidable looking vessel, sirens blaring and all, as we entered the western harbour. They were not happy with us being there, but eventually they allowed us to continue. We finished the circumnavigation in about 12 hours this time.

Martin Reynolds, the Yacht Club rowing captain, arranged a fundraiser for me. He invited *Yantu* to participate in the Yacht Club's annual Hong Kong Henley Regatta (not to be confused with the less prestigious London-based event), which involves some of the territory's best oarsmen competing in several Olympic-class events, such as rowing backwards blindfolded and the famous Sewage Sculls

Race. Well, maybe they were not the territory's best oarsmen and the events not Olympic class, but several strawberries and glasses of Pimms later, that did not seem to matter much....

I was very pleased to be invited, although I was a bit hurt when *Yantu* was volunteered for the Sewage Sculls Race! Teams of two would row *Yantu* around a mark and before returning they would have to cook an egg on the stove. The winner would be the team with the fastest time and 'the highest display of culinary skill'. It was really interesting to see how difficult the oarsmen found it to row *Yantu*. It was clear that race sculling and ocean rowing were two very different disciplines. We had a great day and managed to raise close to US$1,000 for scholarships.

Cash donations had started arriving on a more regular basis from individuals and corporations. More and more logos made their way onto *Yantu*. By the end of July we had raised US$41,000 in scholarships and US$7,000 for race costs. The US$41,000 was 88 percent of a scholarship to send a student to Atlantic College for two years so I called Malcolm McKenzie and told him to go ahead and select a student for the September intake. I also promised him that the remaining US$5,500 would be raised by the end of August. Malcolm was delighted. He had until then been unable to secure funding for an excellent student, Jin Zhou, an 18-year-old male from Guangdong Experimental Middle School. We had sent the first Mainland Chinese Student to Atlantic College!

July 27th was my last day of work as a consultant. I had had a great year working for a great client, so it was a bit sad to pack it all in, but it was either that or not rowing the Atlantic, which was not an option. I was now preparing for the race full time. Véronique was visiting during my last week of work, in Singapore, and we celebrated the occasion together. She then went back to Europe and I returned to Hong Kong to wait for Sun Haibin.

Leaving Hong Kong

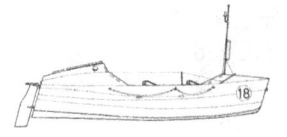

Sun haibin's charter tour arrived in Hong Kong on 1st August. He left his tour group at the airport and met me at the Yacht Club. We were both aware that we did not have much time available, and we went to work straight away.

Long trips out to sea were a priority. We therefore set out on a 21-hour journey from the Yacht Club in Causeway Bay, around the Ninepins, Waglan Island, Middle Island, and then back to Causeway Bay, approximately 30 nautical miles. That was quite a tough trip because we did not manage to get into a sleep routine. We were rowing in two-hour shifts and then went to rest and even though we were exhausted from rowing, we were too excited to fall asleep and got more and more tired. During this trip we also tried the food Maggie Page had sent us. The sheer volume was staggering. There were four meal bags (breakfast, lunch, dinner, night meal). We should eat one meal every six hours. There were an additional two 'day bags' full of sweets, which were for carbo loading. These we were to stuff ourselves with in between the main meals. We tried to eat it all, but it was impossible. We called Maggie and asked her if she was sure the volumes were right. She was absolutely positive. We could not imagine we would ever be hungry enough to eat so much food every day, but who were we to challenge an experienced nutritionist? We told her our final taste requirements and ordered the food. She was pretty certain she could arrange for it to be transported from the UK to Tenerife in the container of one of the British race entries. That was a great relief. It seemed a bit pointless to ship food halfway around the world from the UK to Hong Kong only to ship it out again. And the heat in Hong Kong would probably spoil it.

A few days later we set out on a longer 53-hour trip, covering close to 100 nautical miles. This time we had no problem with sleeping off watch! We started from the Yacht Club and went out around the Ninepins, from where we continued north to Mirs Bay and rounded Ping Chau. We then went east and rounded the small island of Qing Zhou. There we met a number of fishermen who could not quite believe we had rowed there from Causeway Bay. They were very friendly and offered us some fish.

We rowed back towards the Ninepins and I went off watch. When I came back on watch we had not moved much as a strong tide was running against us. I took over and for the next two hours we did not get any further. Then the wind picked up and was blowing a nice force 4, but unfortunately from the south-west, the direction we were heading. Things were getting a bit tricky. We faced the dangers of drifting onto land and wrecking the boat or being carried up into Mainland Chinese waters. If we were spotted there by the coastguard we would be in trouble. We had not brought our travel documents and, who knew, we might get treated as potential drug smugglers and have *Yantu* confiscated!

Neither of these options seemed that good so we both got on the oars and rowed together for the next four hours and got absolutely nowhere. Now it was dark, there was a bit of a sea running, and we were very tired and the rocky coastline looked black and ugly. We decided to empty our excess water tanks. This made *Yantu* about 240 kg lighter, but still no progress. Then we decided to try our sea anchor, but the current was running in the direction of the land and we were approaching it at 0.8 knots. Not good either. Finally, we had to deploy a proper anchor and wait for five hours until the wind and current weakened so much that we could row again. Those were tense hours as we were on the edge of a shipping lane. We then rowed back around the Ninepins and Waglan Island and back into Causeway Bay. Our backsides were slightly sore, but at the time we did not pay much attention to this.

The trip taught us that there was no point in battling against the elements! If the weather was against us, it was much better to take time out and rest instead. We also learned that Hong Kong's waters still have some life left in them. We saw a dolphin and rowed through phosphorescent plankton at night around Waglan Island.

Having completed these trips, we turned our attention to safety training. We rowed to Clear Water Bay Marina where the water was less polluted than in Victoria Harbour. Once there we got some marina staff and residents to help us capsize *Yantu*. It was a good exercise and we could confirm the designers' claim that the design was self-righting. *Yantu* swung back up beautifully!

Sun Haibin and I stood on the side of *Yantu* to help her capsize while the helpers on the jetty pulled on a rope slung below the hull and secured on the far side, which we were standing on. As *Yantu* went over we fell out of the boat. We climbed back in. I opened the hatch to our sleeping quarters to confirm it had remained dry. It had not! About a cubic metre of water was in there. Major shit, to say the least. The electrical panel was completely soaked. How did all that water get in? I climbed inside the cabin and we capsized the boat again. The vents were the answer. Water was pouring in despite them being closed. At the rate it was coming in, I would not be surprised to see a few dolphins coming through, too. I estimate that if we had been rolled three or four times down a big wave in the Atlantic, there would be so much water in the cabin that *Yantu* would not self-right and we would be swimming with the fishes. Thank God we had found this safety hazard! We rowed back to the Yacht Club and they spent the rest of the week cleaning the electrical panel, putting a rubber seal into the vents and tying down the marine battery, which had also come loose during the capsize.

Capsize drills at Clear Water Bay Marina. The boat was self-righting, as it should be, but our cabin vents were not waterproof, resulting in the cabin flooding. That got fixed!

Next it was time to practise sea survival. Together with VIKING we had arranged a safety demonstration for the press on 10th August. As on our previous press days it was raining, but this time a fair number of journalists showed up. We rowed out to Junk Bay and cast anchor as the wind was blowing quite strongly. Here we donned survival suits and life-jackets. We then inflated a life-raft and jumped in. This was very good practice and as a further bonus we made the evening news and both the English and Chinese papers.

Testing our survival suits, life jackets and life raft. The life raft we used for the drill was bigger than the one we brough for the race.

Sun Haibin was leaving in two days time and it was time to focus on the first-aid kit which VIKING had sponsored. The content was prescribed by the race rules and was all Western medicine. We looked at it in disbelief. There seemed to be so much stuff that we could easily have opened a pharmacy. Brian Walker, my long-time GP in Hong Kong, had offered to be our race doctor. Early in the morning we got together at my place for a crash course in first aid. He had brought his Chinese girlfriend Jessica and we set to work. Apart from not knowing what pills and ointments did what, there was another more immediate problem. The usage and dosage directions were all in English and Sun Haibin did not read English. That put me, not Sun Haibin, at a significant disadvantage. I would be able to read the labels and give him first aid, but he would not be able to return the favour.

We therefore went over every single item in the medical kit and Sun Haibin wrote on it what it did. We then divided the medicine in different piles according to usage and put them in zip bags. We then wrote on each bag what it contained, such as 'skin problems', 'digestion', 'sore throat', 'pain relief', 'bandages', 'sewing kit', in both English and Chinese. Brian also helped us prepare a box with the medicines we would most likely need, to keep handy in the cabin. This mainly consisted of band-aids, pain relief pills and ointment, seasickness tablets, re-hydration fluids, and skin lotions. We named one bag the 'Oh, shit!' bag. It contained a syringe and a vial that was to be injected immediately in case of an allergic reaction to prevent loss of life.

Next Brian talked us through stitching up a wound. We knew one team had needed to do this in the last race, so we were paying attention. We then went over CPR, sunstroke, broken bones, and received a few hard tips.

'If your partner is not breathing because he has been hit on the head by an oar or a torn-away solar panel, don't bother to try and resuscitate him. It will only make you feel bad,' Brian told us. Sober news. He continued: 'However, if he is not breathing because of drowning, then there is every reason to resuscitate.'

Brian agreed that he would keep his mobile phone on for the duration of the race so in case we had an emergency we could call him from our Iridium phone. In case Sun Haibin called we would have to hope that Jessica was close by to translate. The training lasted 1.5 days. At the end of the training I asked if Brian could issue us a doctor's certificate to the effect that 'the Applicant is both physically and mentally fit to take part in the Race' as Race Rule 10.3 required. The Challenge Business obviously required this for risk mitigation, but talk about an impossible document to request! 'Could anyone about to row the Atlantic be considered mentally fit?' we joked.

The last thing we did before Sun Haibin rejoined his charter tour and went back to Beijing was to agree on what personal effects we were going to bring. Although we knew we had no chance of winning the race, we still wanted to finish among the top 10 in 56 days and weight was therefore important. Rob Hamill had said in his book that he reckoned the reason they won by such a big margin was purely a weight issue. We had taken this into consideration and *Yantu* had been built and equipped as lightweight as possible. Eventually we agreed on a list:

CLOTHES — EACH PERSON

2 * pairs of thick socks (to avoid blisters from the rowing shoes)
1 * pair of track suit long tights (to block the sun)
1 * boxer shorts (to wear in the cabin)
2 * rowing shorts (one of which was a thick Yacht Club branded pair, which we would start the race in)
2 * T-shirts (one of which was a thick Yacht Club branded pair, which we would start the race in)
2 * tight blouses (to block the sun while rowing during the day)
2 * pairs of gloves (in case of blisters and to block the sun)
1 * sun hat with wide rim
1 * sunglasses
1 * waterproof watch
1 * large towel

1 * small towel
1 * rain jacket

CLOTHES — SHARED
1 * fleece jacket (to keep warm)
1 * sports trousers (to keep warm)
1 * rowing blouse (to keep warm)

DAILY USE — EACH PERSON
1 * tooth brush
1 * razor

DAILY USE — SHARED
1 * tube of tooth paste
1 * bottle of shampoo
6 * bottles of sea soap (which could double for washing clothes, shaving and doing the dishes)
6 * bottles of sun tan lotion
240 * paper tissues (2 people, 60 days at sea, 1 toilet visit per day, 2 tissues per visit)
5 * clothes pegs (for drying clothes)

LUXURY — EACH PERSON
2 * books
2 * cassette tapes
1 * personal diary
1 * English-Chinese dictionary (in case we needed to discuss something complex)
1 * pair of glasses (only for me)
6 * small food parcels with Chinese-style food (only for Sun Haibin)

LUXURY — SHARED
1 * Walkman
1 * shortwave radio
1 * camcorder with spare tapes and batteries
2 * disposable waterproof cameras
1 * fishing kit

Apart from that list, we agreed not to bring any other personal or luxury items. In the end this turned out to be way too much stuff, but at the time it seemed very little!

Leaving Hong Kong

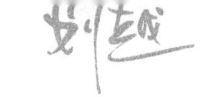

We both felt we had achieved a lot in a short time and, on the 12th August, Sun Haibin went back to Beijing to work on getting his visas. I turned to the task of transporting *Yantu* to Tenerife in time for the race start.

I had not been able to find a sponsor for the transportation and was therefore looking around for the best commercial alternative. The Challenge Business' suggested shipper was not competitive and in the end I ended up using Maersk. I felt somewhat disappointed that it was not willing to sponsor the transport as it is a Danish company and it had just spent millions of US dollars to help China win the right to host the 2008 Olympic Games.

Timing was important. I wanted *Yantu* delivered on the pier on 26th September in Playa San Juan, Tenerife, so that we would have 10 days to prepare before the race start on 7th October. To make this schedule *Yantu* needed to be packed and shipped out of Hong Kong on 23rd August. Maersk came down to the Yacht Club to measure *Yantu*, which was very tall as she was sitting on her makeshift cradle. She was too long to fit into a 20-foot container. I therefore had to ship a 23-foot boat in a 40-foot container, which seemed a waste and was more expensive.

While investigating freight possibilities, Kenneth Crutchlow, executive director of the Ocean Rowing Society, got in touch. The society's purpose is to further ocean rowing and to record and validate rows. Kenneth was therefore keen to assist where possible. Based on his experience from the 1997 race, customs clearance in Tenerife was a big problem, particularly for boats arriving from outside the European Union. He therefore suggested that the society, which had contacts in Tenerife, should get *Yantu* through customs and deliver her in Playa San Juan. It would be US$300 more expensive than Maersk and Sun Haibin and I would have to pay membership fees to join the Ocean Rowing Society, but I decided to go for safety and let Kenneth handle procedures in Tenerife.

About this time Dominic Biggs contacted me. I had read about Dom, but never met him. He was working as a journalist for the *South China Morning Post* in Hong Kong and was also in the race together with his best friend Jonathan Gornall, who was a journalist at *The Times* in the UK. A year ago, before *Yantu* started getting media coverage, he had written an article in the *Post* outlining his upcoming entry in the 2001 Atlantic Rowing Race, stressing he was the only Hong Kong resident to participate. I had been a bit disappointed about this, as I knew that Chris Perry had told Dom that I existed, but Dom was running his own campaign and was of course trying to make it as unique as possible. He had not written any more articles, but every time the *Post* reported on us he would also get a mention.

Dom's boat was with his partner Jon in the UK, so he had not had much practice rowing it. I therefore invited him to come along to Clear Water Bay Marina for another capsize drill to verify that the vents were now watertight. We started rowing and talked about the upcoming race. Jon had already arrived in Tenerife and reported things were chaotic there. In 1997 the race had started from the harbour of Los Gigantes and the Challenge Business had announced that we would start from there as well. Unfortunately, they had forgotten to ask the Los Gigantes harbourmaster, who learned that the race would start from his harbour by reading the local newspaper. Obviously, he was not impressed and had told the Challenge Business in no uncertain terms that there was no way he would allow the race to start from Los Gigantes. The Challenge Business insisted it would and things were tense in Tenerife. Jon, being the first rower to arrive, was getting the brunt of this. Having exhausted this subject we then turned to the challenge of teamwork.

Dom and Jon had been friends for over 20 years. Dom thought Jon was a bit of a character and highly strung. Jon's ambition was to win the race. Dom did not think they could do that and just wanted to make it across in a decent time. This was a bit of a sticking point between the two, but Dom was confident that they would sort it out. He told me more about Jon and eventually I asked him if he was sure he really wanted to row the Atlantic with Jon.

'I can't imagine doing it with anyone else,' Dom replied.

We continued rowing in silence. I thought their setup sounded explosive, but said nothing. Then I thought of Sun Haibin and myself, two completely different cultures. Were we aligned? Could we go the distance? The more I thought about it, the happier I became. I began thinking that the fact we came from two completely different cultures and backgrounds was an advantage. Because we had not known each other for very long, we would not take things for granted or leave things to chance.

Leaving Hong Kong

We wanted to do the race for personal achievement and to showcase international understanding and cooperation. We could only succeed in doing this if we made it all the way across and were still friends on arrival. Our goal was to finish within the top 10 in 56 days, but if that was unrealistic we would still row on and make it across. Not completing the race would only be acceptable due to a safety issue. We knew we were bound to annoy each other, because of our vastly different cultural and social backgrounds, and we had therefore prepared ourselves mentally to cut the other person a lot of slack. If we were forced to give up we would give up as a team and bear the brunt of failure together as a team. I would be the captain and Sun Haibin the crew because I had the most sea experience and navigational skills. All this we had already discussed at length and agreed. Nothing was left to be decided out at sea. It seemed our big cultural differences might be working in our favour! We were aligned on our objectives!

Dom and I arrived at Clear Water Bay Marina and conducted the capsize drill. To my relief, the vents did not leak and the cabin stayed dry. We rowed back to the Yacht Club.

The days immediately before shipping out *Yantu* were spent trying to install the water-maker. From the experience of the 1997 race rowers, I knew that desalination failure had been the single biggest cause of teams having to retire from the race. Fresh water was critical. We would need about 12 litres a day each, 10 litres for drinking and the rest for cooking and personal hygiene. This meant that every day we would be consuming 24 litres of fresh water. If we brought it all instead of manufacturing as we went along, we would need about 1.4 tons of water, which was clearly not feasible. If the water-maker failed we could use the 150 litres of emergency fresh water that the race rules stipulated as compulsory, and which made the boat self-righting in case of capsize. At the rate of 24 litres per day that would last six days, long enough for a rescue vessel to get to us, but not long enough to complete the row. Moreover, if we touched this water, we would be disqualified from the race. Thus the water-maker was *very* important!

The recommended water-maker for the race was a PUR 40E. Mechanically, it was simple enough. An electric pump, driven off our marine battery, which in turn was charged by the solar panels, would move a piston which would press salt water through a delicate membrane that would withhold the salt crystals and dispose of them through a waste water pipe. Fresh water would trickle through the membrane at a rated output of 4.5 litres per hour. We would collect this in three 10-litre jerry cans in the cockpit, which would be used and replenished on a daily basis. In case of electrical failure the water-maker could be operated manually.

We tested the water-maker at the Yacht Club. After an hour the jerry can should have been half full, but it was hardly wet on the bottom. Crisis! Ah Bun, the head mechanic at the Yacht Club, and I took the water-maker apart and put it back together. It still did not produce water. More people arrived and helped out, but no matter what we did, we could not get it to work, even in manual mode. There was nothing to do except send the unit back to the supplier and get them to service it. We would have to install the water-maker in Tenerife when we got there. If we could not get it to work then we would be out of the race before it had even started!

I went to look for Duthie to discuss the best route across the Atlantic. Poring over Admiralty Chart 4012 of the North Atlantic Ocean (Southern part) and the corresponding routing chart we discussed strategies. In 1997, the winners had rowed straight across. At approximately 2,500 nautical miles this was the shortest distance, but the chart showed significantly higher chances for adverse weather than if we first rowed south until we were below 21 degrees latitude before rowing west. We would have to row about 200 nautical miles further, but in all probability we would have better weather. If everyone else went straight across we would be in different weather systems from the rest of the fleet, which could be an advantage. If the fleet north of us got a big dose of adverse weather we could pick up a few positions rowing in fairer conditions. We reckoned we would row slightly faster than two knots and did the maths. The total distance of 2,700 nautical miles divided by 48 miles per day meant we should get across in 56 days, which was our target. We decided to go south and, like a great artist, Duthie drew an arc on the chart with a pencil. I looked at the map, which was about a metre wide. In the margin on the left I could see a bit of South America and on the right a bit of Africa. Everything in between was water and the numbers on the chart showed depths of about 4,000 metres. Duthie's thin pencil line on the chart was quite sobering, but at least we would not have to worry about running aground!

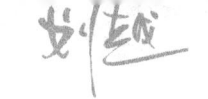

As we were rolling up the charts Duthie said jokingly: 'You know, on race day something completely unexpected is bound to happen and all our planning will go to shit!'

The day before *Yantu* was due to be shipped out, a new bombshell exploded on the project. Some time after *Yantu* arrived in Hong Kong I had received the invoice for assembling and painting the hull from Luyang Boat Building Company. One look at it was sufficient to determine it was not 'at cost'. I had written a detailed reply and made a counter offer, but did not hear anything back. Wanting to settle the invoice I had subsequently made another slightly higher offer and Luyang replied with a figure slightly lower than their original invoice. We were still many thousands of dollars away from an agreement and had had a lot of heated communication, but it was not going anywhere.

I got a call from Mr Guo: 'Are you going to pay the invoice?'

'I am eager to pay the invoice, but not the amount you propose. It is not a fair cost price,' I replied.

'If you don't pay the invoice today I will put a writ on your boat and prevent it from leaving,' Mr Guo threatened. I was furious.

'If you do that I will tell all my media contacts, supporters and sponsors that the only reason we are not participating in the race is because of you,' I countered, refusing to back down.

'Fine,' Mr Guo said and hung up.

Shit! Maybe he was bluffing, but I couldn't be sure. I called a lawyer friend of mine and explained my situation. She told me that she had to call me back as she specialised in common law and this was a marine law issue. Ten long minutes later she called back. The news was not good. Mr Guo could indeed legally stop *Yantu* from leaving Hong Kong until the invoice was paid if he decided to do so. What do to? There was less than 24 hours until *Yantu* should ship out. Would there be time to put a writ on *Yantu*? I did not know, but I knew one thing: there was no way I would pay the amount he demanded. We had entered into an agreement based on goodwill and I was not going to be held hostage. I decided to continue working on *Yantu* and prepare for the send-off press conference the next day, but things seemed impossible to manage alone. I called Rickie Tsui, fellow UWC alumna, who had contacted me out of the blue some time before to see whether the Yantu Project was for real.

'Rickie, didn't you tell me you always wanted to organise a press conference?' I asked.

'Yaahhh,' the reply came back.

'Well, we are having the *Yantu* send-off press conference tomorrow. Are you game?' I continued.

'Sure!' Rickie replied, ready as always to lend a hand.

One hour later he showed up at the Yacht Club, I introduced him to Ellen Wong and they went to work. I returned to *Yantu*, where the sleeping cabin still needed to be made comfortable. Sun Haibin and I had decided that two layers of camping mat would be sufficient to sleep on and we had bought some pipe insulation to wrap around the supporting beams in the cabin so that we would not hurt ourselves too much if were thrown around inside the cabin in bad weather. I was gluing down the padding when Keith arrived.

'Mr Guo called me and told me he is going to put a writ on your boat. I talked to him and I have the following proposal. You deposit the money Mr Guo requests into an escrow account. We then get a qualified marine surveyor to determine the fair cost price. His judgement will be final. You will have to pay half of Mr Guo's legal fees as well as half of the marine surveyor's costs. What do you think? Keith asked.

'Will it enable *Yantu* to be shipped out on time?' I asked.

'Yes, but the whole thing has to be completed today,' Keith replied.

I thought about it. The clock was ticking and Mr Guo and I were way too entrenched to find another solution in the available time.

'OK,' I said.

Covered in paint and glue stains I headed off to the expensive part of Hong Kong's business district to find a lawyer to represent me. Not exactly how I had pictured spending the last day before *Yantu* was shipped out, but by the end of the day the escrow agreement was drawn up, the disputed sum of money deposited and the marine surveyor booked for the following morning. *Yantu* was going to leave after all! After returning from the race I ended up paying almost US$5,000 less than Mr Guo's original invoice, even after having paid half of his legal fees and surveyor costs. The tense relationship with

Leaving Hong Kong

Mr Guo is my single biggest regret from the race because I am convinced he is a nice and progressive man. Had we struck a different deal where he had branding on the hull, we could have rowed *Yantu* from Hong Kong up to his factory after the race as part of our homecoming and called an amazing PRC press conference for both Luyang and *Yantu*. But that was not to be. There is a saying in Chinese: 'Let out a long line and you will catch a big fish'. We had both fished with too short lines.

The *Yantu* send-off party, or Bisan, as the Yacht Club called it, was great. Another suckling pig was served up along with beer and soft drinks. Then Robert Bird, the general manager of the Yacht Club, assisted me with burning incense for good luck.

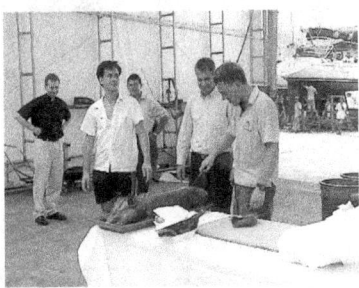

Bisan ceremony asking for the sea goddess Tin Hou for good luck. Burning incense with Robert Bird, the General Manager of RHKYC and cutting the suckling pig.

The majority of the active supporters and sponsors had shown up as well as the Yacht Club staff who had helped getting *Yantu* ready, in total about 50 people. I had to laugh at myself. Three years before, when I had signed up for the race, I had thought that building an ocean rowing boat and rowing it across the Atlantic was going to be a piece of cake. I could not have been more wrong! It had been a huge undertaking to get ready and without the support of all these people I would never have got there! I did a number of press interviews and really regretted that Sun Haibin could not be there to share in the celebration. It was a fantastic day!

Shipping out boat day in Hong Kong. Without this big support crew, we would never have got to the starting line. Notice that Sun Haibin is missing. He was in Beijing trying to sort his travel documents for going abroad.

Sticker where the Boatyard crew signed their names to wish us good luck. It so happened that the "YOU CAN" was lit up at night by the light from the compass above the sticker. This reminder helped to keep us going.

Leaving Hong Kong

My friend Kate's husband, Matt, the Church of England vicar, came to the Bisan in full regalia to wish us luck and bless *Yantu*. While Sun Haibin was still in Hong Kong another friend of mine, Sandra Gonzalez, who works as a healer, had also blessed the boat. She had burned white sage inside the cabins, used by the North American Indians to cleanse and purify the environment and dispel negative forces. She had also given us three crystals to bring along: Amethyst for healing and connecting with God; rose quartz to enhance respect and harmonise communication and solidarity in the boat, and finally clear quartz to enhance our clarity and sense of purpose. We were told to wash them from time to time to keep them strong.

Sandra had also worked with us on meditation techniques to combat fear. She maintained that fear and excitement release the same body chemicals and are very close in vibration. Therefore, whether a situation would cause us to need to change underwear or give us the ride of our life would mainly be a mind game, she explained. By the time *Yantu* was shipped out it had been blessed a total of four times covering a wide array of faiths. *That can't do any harm,* I thought.

Staff at the RHKYC putting Yantu into her container ready for shipment. Next stop Tenerife!

Beijing to Barbados in a Rowboat

Before leaving, *Yantu* had to be put into the container. Despite previous measurements she turned out to be too tall because of the cradle. The boatyard got out welding equipment and quickly fixed this. The bow now pointed slightly upwards and it almost looked like she was sitting on a launch ramp. I watched the truck with the container leave the Yacht Club. It was finally happening! Relieved, I went home and crashed. The last few days had been extremely intense.

The next day I went back to the club to meet with Robert Bird to settle my accounts. We looked at the amount of sponsorship money on account. The scholarship account stood at close to US$49,600, US$3,100 more than what was needed for Jin Zhou's scholarship. Great! I could fulfil my promise to Malcolm. I arranged that the Yacht Club would forward a cheque to Atlantic College to clear the scholarship account. We then turned to the race-cost account, which had raised US$19,600, but most of this was already spent. There was not enough left to pay my bill to the Yacht Club, but money kept coming in both for scholarships and race costs, so maybe there would be enough money by the time we got back from the race.

'Do you want me to pay now or when I come back?' I asked Robert.

Silence.

'Maybe it is better if you pay now, just in case,' Robert finally said. We both laughed and I wrote him a cheque.

I never doubted that Sun Haibin and I would come back alive. We had put so much effort into the project and it was for a good purpose. We simply deserved to succeed!

But before we could come back alive, we first needed to go and Sun Haibin still did not have his visas for either Tenerife or Barbados! He had submitted all the documents, but when he phoned to enquire, the reply was always that it was 'being processed'. In the end, I decided to fly to Beijing and accompany him to the embassies. Chances were that if a Westerner was sitting in the visa application waiting area, someone would notice and we could get straight answers. It worked and finally he got his visas with a full two days to spare before we flew to Tenerife, so absolutely no stress....

Leaving Hong Kong

The cross-cultural ocean rowing duo about to fly to Tenerife.

Tenerife — meeting the competition

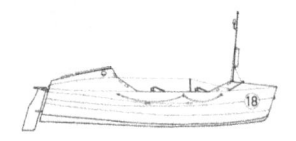

We flew from Beijing to Hong Kong on 15th September, then left for London Heathrow on 18th September (lucky date!). Sun Haibin was allowed a three-day stopover in Hong Kong because he had an ongoing ticket, yet another peculiar PRC travel regulation. Because he had needed an air ticket to apply for his visas we had ruled out the possibility of holding out to find an airline to sponsor us and had purchased the tickets early on. I had spent the last year travelling all over Asia for three weeks per month and had therefore accumulated significant airmiles, so we got upgraded to business class. It seemed like a good omen for things to come. Sun Haibin had only flown domestic planes in China and could not quite get over the luxury of Cathay Pacific's business class.

We arrived in London the following day. Due to the terrorist attacks of 11th September, everything was delayed and security was tight. We were not allowed to check our baggage through and in London it needed to go through security again. Carrying a bulky reverse osmosis water-maker unit, a VHF marine radio, an Iridium satellite phone, a camcorder, and two handheld GPSs, we did not look like ordinary travellers and were taken off to have our baggage searched carefully, but eventually we were cleared for the flight to Tenerife where we arrived jetlagged later that same afternoon.

We rented a car, looked at a map and headed for our hotel on the south side of the island. The hotel was close to Los Gigantes, where the race office was and where we would start from if the Challenge Business and harbourmaster could reach an agreement. The Challenge Business had promised we would be able to get water berths in Los Gigantes harbour four days before the race.

Tenerife — meeting the competition

Playa San Juan was a small and sleepy fishing village 20 minutes drive from Los Gigantes, with a small harbour protected from the force of the Atlantic Ocean by a wide breakwater that we could drive out onto. Until we could get into Los Gigantes the boats would be kept here on their trailers and lowered into the water either by the slipway or a small crane used for lifting the daily catch out of the fishing boats.

One boat had already arrived and we went to have a look. It was called *Mrs D* and was rowed by brothers Steven and Mick Dawson. They told us that they had also stayed in our hotel, but had moved to Playa San Juan because it was much more convenient. *Good idea,* we thought, and went to look for a hotel room. When Véronique and later Bettina Tria, a good friend of mine from Atlantic College who wanted to come to the race start to support Véronique, arrived, our current hotel room would become cramped.

We found a nice apartment overlooking the ocean and without bargaining we got it for slightly more than half the listed price. We then went back to check out of our hotel, having confirmed first that the UK travel company with whom we had prepaid the stay would reimburse us. In order to simplify the refund I paid for the night we had stayed and was charged half the price we had pre-paid to the travel company, who had assured us they had negotiated a special Ocean Rower discount. *Some discount,* I thought. We drove back to Playa San Juan and settled into our spacious self-catering apartment.

Life in Playa San Juan revolved around a bakery and a pub with an outside seating area consisting of a few plastic chairs and tables directly opposite the entrance to the breakwater. We bought some bread and sat down to enjoy our lunch.

'You must be Christian and Sun Haibin!' a larger-than-life voice proclaimed from a few tables away. The voice belonged to a big man with long greasy hair sticking out from under an old baseball cap. He stood up, picked up his beer, and came over to sit down.

'Kenneth Crutchlow, Ocean Rowing Society. Great to see you!' he said and sat down shaking our hands. I was really pleased to meet Kenneth, who had enthusiastically been giving us advice for the past year or so. We talked for a while and then he said: 'You know, I forgot to charge you for the crane to unload from the container. That will be an extra US$65.'

'Has our boat arrived?' I asked.

'I think so. Why don't we drive to the customs area in Santa Cruz to have a look,' Kenneth replied. I translated for Sun Haibin and asked him what he thought. It was a 40 minute drive, which was not particularly appealing but, like me, Sun Haibin was keen to see that *Yantu* had actually arrived so we all piled into our rental car and set off. We spent the rest of the day escorting Kenneth while he ran errands to do with other rowing boats. In fact *Yantu* was nowhere to be seen. We had paid the Ocean Rowing Society quite a lot more than Maersk had quoted for delivering *Yantu* in Playa San Juan and did not expect to have to help Kenneth run errands. Late afternoon we arrived back at the plastic chairs in Playa San Juan, none the wiser about *Yantu's* whereabouts. A complete waste of a day! Sun Haibin and I discussed what to do.

'Kenneth, you don't need us for customs clearing as long as you have our passports. We will go and get them,' I said.

Kenneth resisted. 'I am not really comfortable with keeping your passports. What if I lose them?'

'I am sure that won't happen. We'll go get them,' I insisted, wondering what the big deal was.

'OK, but if you are not coming back with me tomorrow I will have to borrow your rental car,' he proclaimed. I couldn't understand why he expected us to drive him everywhere and why he didn't rent a car himself. Eventually Kenneth got the Dutch team, Rik Knoop and Michael Tuijn, to drive him.

Staying above the bakery was the American team and their support crew. At 41 and 51 respectively, Tom Mailhot and John Ziegler were among the oldest oarsmen in the race, but it did not show. They were big fit blokes and built like tanks. They had already been in Playa San Juan for a while and were waiting for their boat, *American Star*, to clear customs. We struck up a conversation and immediately hit it off. Despite not being able to communicate verbally, Tom and Sun Haibin had clearly clicked and were having a great time. Tom was a carpenter by profession, although the Challenge Business had put him down as a 'Professional Adventurer', which he found embarrassing, and provided us with the perfect opportunity to give him endless grief. The title was not necessarily that far off, though, as Tom had kayaked around Cape Horn. John was a food wholesaler by profession and an avid Hawaiian outrigger competitor. In 1966 he had read about Chay Blyth and John Ridgway's row across the Atlantic and had planned to commission his own ocean rowing boat to have a go at it himself. The project had not come to fruition and it had gone on the back burner until he became aware of the 2001 Atlantic Rowing race. This time he had to do it! Tom and he were there to have a good crack at winning, and John's determination had cost him his marriage. They were running each morning for an hour and asked us if we wanted to come along. We were happy to join in and we continued running together every morning until the race start. Despite their big frames, Tom and John were strong runners. The pace suited me, but Sun Haibin was always off in front and hardly sweated on return. Using hand signals, Sun Haibin was teasing Tom that he was an old man and should really run faster. 'Damn you, Sun Haibin. You wait until my girlfriend gets here and we will see how fast you really are,' Tom would reply.

Tom's girlfriend, Sarah Evertson, and Véronique arrived around the same time. All of us quickly got on well. Sarah and Véronique were glad to finally find someone who could share the stress of having a boyfriend rowing across the Atlantic. Sarah joined us on the morning runs, but despite her being an experienced marathon runner, Sun Haibin still got away. Tom trying to outrun Sun Haibin became a daily joke. He would watch for a time when Sun Haibin was not paying attention and sprint off to try and leave him in the dust, but to no avail. One day Tom asked Sun Haibin to write some Chinese characters on his sun hat for good luck. Sun Haibin readily complied and wrote 强健, , meaning 'Strong health'.

More and more rowers and supporters were arriving in Playa San Juan and the area outside the bakery was buzzing from early morning to late evening. It was great to be surrounded by people who were all going to row the Atlantic. In Hong Kong we had been oddballs and were frequently asked questions like: 'Aren't you afraid you will die?', 'What if a shark attacks the boat?', 'Do you think the boat is strong enough to weather a hurricane — have you seen *The Perfect Storm?*' and other similarly helpful questions. We had even been told flat out that we could not do it and would die trying.

The human race is funny like that. Every time something out of the ordinary is attempted, people will go to great lengths to convince you that you cannot succeed, regardless of whether they have any knowledge of the subject. Very few people embrace new ideas with a 'sounds great, go for it' attitude. Yet it is only when people break the mould that significant progress is achieved, either personal or historic. Ira and George Gershwin wrote a superb song in 1937 playing on this peculiar tendency of human nature to look for the negative. Part of it goes:

> *They all laughed at Christopher Columbus*
> *When he said the world was round.*
> *They all laughed when Edison recorded sound.*
>
> *They all laughed at Wilbur and his brother*
> *When they said man could fly.*
> *They told Marconi, wireless was a phony*
> *It's the same old cry.*
>
> *They all laughed at Rockefeller Center*
> *Now they're fighting to get in.*
> *They all laughed at Whitney and his cotton gin.*
>
> *They all laughed at Fulton and his steamboat*
> *Hershey and his chocolate bar*
> *Ford and his Lizzy, kept the laughers busy*
> *That's how people are.*

We only live once — actually, that is not completely true. A lot of people exist, but do not actively *live* their lives. We have a maximum of about 100 years on this planet, so why live a life you are not happy with when you can pack it with fun, excitement and fulfilment by putting yourself in the driver's seat?

This does not mean I advocate pursuing any crazy idea in a reckless manner. It took a lot of research and testing to put a man in space, but it was achieved against the odds because of 'can do' attitudes. It is important to prepare for an adventure properly, but when all is tested and you still believe it is possible, then why not go for it regardless of what others think? That creates personal or bigger breakthroughs.

I think a key point about adventuring is often misunderstood. No one in his right mind does an adventure to die. You do it to live! When you are in the middle of the elements and pushing your limits you feel the risk and you feel alive!

Everyone who showed up in Playa San Juan to row across the Atlantic had evaluated the risks and come to terms with them. Instead of focusing on the negative, conversations were centred around the

Tenerife — meeting the competition

best route, the water-maker installation, who would win and so on. Here anyone who was not rowing across the Atlantic was the odd one out and it felt great!

The number of boats on the breakwater started to increase. *Yantu* was still missing, but finally, on Monday 24th September, I got a call from Kenneth that it had arrived. But there was a problem. Maersk had no records that the freight had been paid and would not release it. We had paid both Maersk and Kenneth before *Yantu* had left Hong Kong and I had asked them to coordinate directly to avoid problems. Now I had to place several calls to Maersk in Hong Kong so they could send through the confirmation that the freight had indeed already been paid. Problem solved, or so I thought. Early morning on 27th September, Kenneth called me again. '*Yantu* is not on a trailer,' he said. 'We cannot tow it to Playa San Juan.' If Kenneth had chosen to take the boat out of the container prior to delivery in Playa San Juan that was his problem. The premium price I had paid the Ocean Rowing Society seemed to have been a poor investment.

I lost my temper. 'You deliver my fucking boat and cradle to Playa San Juan,' I said. We argued for a while and Kenneth hung up. *Yantu* arrived later. It was great to see her again and confirm she had not suffered any damage during transport. However, she was without her cradle and had been put on tires 50 metres from the crane. We needed that cradle. When it finally arrived, the other rowers lent a hand and lifted *Yantu* up onto it. We then pushed it down the other end of the breakwater to be close to the crane as we were less mobile than the boats with trailers.

Our spot was third from the end. The two Kiwi entries were to our right. Two girls, Stephanie Brown and Jude Ellis were rowing Rob Hamill's old boat from the 1997 race, now renamed *Telecom Challenge 25*. The boat at the very end was the new Kiwi boat, which would be rowed by Rob Hamill himself and Steve Westlake, called *Telecom Challenge 1*. They were running a campaign fully-sponsored by New Zealand Telecom and the boats had been airlifted from New Zealand. Instead of trailers they had small plywood cradles with tiny wheels, allowing them to fit into the hull of the aircraft. On the left side was *Troika Transatlantic* belonging to husband-and-wife team Debra and Andrew Veal. *Yantu* was sitting in the middle, about one metre higher than our neighbours. Sun Haibin and I climbed into *Yantu*. From our elevated view we started studying our neighbours. The Kiwi side looked very focused and was milling around busying themselves with the boats not even noticing their new neighbours. Debra and Andrew were more relaxed so we talked to them for a while. *Troika Transatlantic* had *WellHungArt.co.uk* written on the side of the cabin and this of course begged a question. It turned out it was Debra's own company, which helped artists promote their work online. Suddenly I thought I was seeing double, but it turned out to be Debra's twin sister. Andrew was a consultant. It had originally been Debra's idea to row the Atlantic and then Andrew was roped in. They looked like a good team.

Helping Andrew and Debra Veal launch. The race brought Debra fame when Andrew asked to be rescued a few days into the race. Undeterred Debra continued on her own, finishing the race in 111 days.

We turned our attention to the Kiwi side again. The two girls both looked incredibly strong and athletic. Steve was tall and also very athletic. I knew from press releases that he was a policeman. In 1997 Rob had rowed across the Atlantic with Steve's police partner Phil Stubbs. I also knew from Rob's book that Phil had tried to boot Rob out of the race and row with Steve instead. It appeared strange that Steve and Rob were now rowing together, but subsequent to the 1997 race, Phil had died in a plane crash and Rob and Steve had now paired up to row together in remembrance of Phil, so in the end it all made sense. They were there to win the race and break their own 41-day record. Having seen the other boats and competitors it was clear that they had a very good chance of doing both. It was almost as if they were preparing for a completely different race. Rob was nowhere to be seen, but finally a shorter and also top-tuned Kiwi athlete appeared on the breakwater with a fit-looking girl. It was Rob and his wife, Rachel, who was a competition swimmer.

The two Kiwi crews. Stephanie Brown and Jude Ellis in the foreground and Steve Westlake at the back, all hard at work and totally focused.

Rob and I knew each other from phone calls so I called out his name. He came over, we shook hands, and I introduced him to Sun Haibin. He in turn introduced us to his teammates, who immediately went back to work again. Rob was quite lively and had a good sense of humour so we were soon giving each other stick.

Sun Haibin and I drove to Los Gigantes to announce our arrival at the race office. Teresa was there. It was great to see her again and I introduced her to Sun Haibin. She looked stressed. The situation with the harbourmaster was not improving and she was taking the brunt of criticism from the rowers. We collected the race crisis operation plan, which included a set of communication guidelines based around the Inmarsat D+ unit, which the Challenge Business provided for race tracking. The unit would be polled once a day and positions updated on the race website so that supporters could see how all the boats were doing. The Challenge Business was using two of its BT Global Challenge boats as safety boats and they would follow the fleet across. Each team was also required to carry an EPIRB (Emergency Position Indicating Radio Beacon) which, when activated, would send an automatic SOS and position to the nearest coastguard, who could then redirect nearby vessels to the rescue or send a plane to spot the boat. In the 1997 race, the brothers Matthew and Edward Boreham rowing *Spirit of Spelthorne* had activated their

EPIRB and after five days they had finally been found. Edward had suffered a nervous breakdown mid-Atlantic and had started swimming for shore. Matthew had managed to get Edward back on board and had set off the beacon. The seriousness of the race was starting to dawn on us. The list of communication abbreviations for the D+ unit also reminded us of this — one of the codes, X, meant 'My rowing partner has died'.

We would only be using the D+ unit to enable the Challenge Business to position poll us, as we had the Iridium Satellite phone, which would enable us to receive and place calls as with a normal mobile phone and therefore was far superior to the D+ unit.

Stratos had only recently taken over the Iridium satellites from Motorola and were offering very good deals to entice subscribers, such as being able to call the Iridium phone at the cost of a normal landline call through a switchboard in the US. Stratos' largest customer was the US army and after the events of 11th September, when the US was preparing to go to war against Afghanistan, the switchboard number in the States ceased to work, probably because they suddenly had a large number of subscribers from the army! As a result, we suddenly could not receive incoming calls, which we had counted on to contain costs. Instead we would have to make all calls at US$1.5 per minute once we had used up our 600 sponsored minutes and we had to change our whole media strategy.

Before leaving Hong Kong, we had arranged that Stratos and the Yacht Club would liaise with journalists, who could then call through to us at fixed times. Since we could no longer receive calls, this setup did not work and we decided to call Hong Kong and PRC journalists directly from the boat instead, as we knew them quite well. In Denmark, my old sailor friend Olav took up the challenge to drum up interest amongst the Danish press.

We had pre-arranged a weekly 10-minute live radio interview with Radio 3 every Saturday at 11AM Hong Kong time for the duration of the race and Victoria Button had also promised to continue writing articles about us in the *South China Morning Post*. We had also promised our sponsors and supporters a weekly update, which Véronique would compile and dispatch by e-mail after talking to me.

I called Victoria to suggest she write an article about our arrival in Tenerife. Much to my surprise she replied she could not do it. 'Why not?' I asked. 'Dom, the other rower from Hong Kong who works here at the *Post* has already written about their arrival in Tenerife and we cannot write the same story twice.' This was a major blow. Over the last year, the *South China Morning Post*, which is the largest English-language paper in Hong Kong, had run regular articles on us. It just did not seem fair that we were going to be dropped simply because Dom had the inside track of working at the newspaper. I told that to Victoria, she went to work and we got back in the paper again.

As soon as we got the boat, we started installing the water-maker. We walked down the line of boats and talked to the other rowers to try and figure out the best installation. Some were in the same situation as us and had yet to install it, others had installed it but did not want to discuss it, but the majority were happy to share experiences. The Dutch team had a long hose with a weight at the end, which they would throw over the side of the boat to draw in seawater. This arrangement avoided getting air into the inlet hose, which was difficult to avoid if a through-hull sea cock was used as the boats did not have more than about 30 cm of draught and easily rolled 45 degrees or more, so air constantly washed under the boats. Simon Walpole and István Hajdu rowing *UniS Voyager* used a sea cock inlet and had built a sophisticated looking pipe-shaped air trap to avoid air getting into the water-maker. Air in the water-maker was what we all had nightmares about. It caused the output to drop significantly or stop altogether and was difficult to expel. It could well mean the end of the race. We dropped in on Alex Wilson and Rory Shannon rowing *Atlantic Warrior*. They had a through-hull sea cock inlet and had turned the standard water pre-filter upside down so it worked as an air trap. The pre-filter was supposed to sieve impurities out of the seawater prior to it entering the water-maker as contamination would also cause the water-maker to cease working. What they had done therefore seemed a bit of a gamble, but they were rowing a boat from the 1997 race where the same installation had worked fine throughout. On reflection the Atlantic Ocean was a lot cleaner than Victoria Harbour so maybe it was not such a gamble after all. We visited the Belgian team Pascal Hanssen and Serge Van Cleve rowing *Bruxelles Challenge*. They had a similar installation to *Atlantic Warrior*, but had added a one-way fuel pump into the intake hose between the sea cock and the pre-filter. They had also added a valve on top of the inverted pre-filter. By opening the valve and using the pump they could expel air in the pre-filter and fill it completely with seawater without having to remove it. When they closed the valve again the seawater did not run back out the sea cock because of the one-way fuel pump. It seemed an excellent idea.

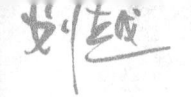

We arrived back at *Yantu* and climbed on board. What had the Kiwis done, we wondered, and peeked down into their boats. Their installation was completely different. Instead of placing the water-maker in one of the side lockers like everyone else, they had placed it in the compartment under the stroke rowing seat together with the 150 litres of emergency fresh water. It looked interesting, but something was wrong. According to the design, the compartment was supposed to have two floorboards, but they were not there. It was not a big deal, but, since they were such a serious bunch, I could not resist stirring them a bit: 'Hi guys, what have you done to the floorboards in the compartment with the water-maker? Did you burn them on the Christmas log fire? According to the race rules weren't we supposed to use all the pieces in the building kit?'

What happened next was straight out of Harry Potter. As if I had uttered a magic spell Steve and the girls stopped what they were doing, came over to *Yantu* and struck up the friendliest conversation of all time. They were obviously a bit concerned about the floorboards themselves. '*Yantu* looks good,' they commented. 'Do you need any help? Do you want to borrow any tools?' Jumping at the chance I said a few hints about water-maker installation would be nice. Steve spent the next half hour dumping knowledge on us. 'Make it as simple as possible,' he advised. 'You don't need all the fitting braces, just secure things with shock cord. It's lighter and makes it easier to get the unit in and out in case you need to. The real secret to the unit working well is to ensure the inlet into the water-maker unit is below the waterline as the unit is gravity-fed.' He subsequently climbed on board and showed us where he would put the unit in the side locker. We thanked him profusely. He climbed down from *Yantu* and rejoined the girls. It did not take them long to return to their original focused selves.

The magic spell invoked by mentioning the missing floorboards provided us with good entertainment for a while, but unfortunately it stopped working after the Kiwis passed scrutineering without a problem. After that nothing could distract them from optimising the performance of their boats.

Tenerife — meeting the competition

Installing the watermaker.

Using Steve's advice and the Belgians' fuel-pump idea we completed the installation in about one hour. It looked good, but did it work? We pushed *Yantu* over to the industrial crane and went to look for the local harbourmaster to operate the crane. *Yantu* was lowered into the water, we got in and rowed out into the open ocean. It was a great feeling to be rowing again and for a while we just enjoyed gliding along. I dropped out from rowing and started priming the inverted pre-filter *cum* air trap with the hand pump. I switched on the water-maker. The pump started working and waste water came out. We waited in silence. After a while water started dripping out of the freshwater hose. One taste confirmed it was drinkable. It actually tasted rather good. We high-fived. I stuck the freshwater hose into a jerry can, noted the time and went back to rowing. One hour later we checked the jerry can. It was half full! The water-maker was working! After an impromptu water fight we rowed back into the harbour with huge grins on our faces. What a relief!

I climbed up onto the breakwater and went to look for the harbourmaster while Sun Haibin stayed in *Yantu* and kept her from banging into the pier. The hoist was adventurous as the crane was perhaps a bit small for the boat, but we managed to get back up. We were set down rather hard on the cradle and had damage to the skeg. This became a constant problem each time we were pulled out of the water as the crane was a bit rough in its manoeuvring. By the end we were quite competent in epoxy repairs!

Having got the water sorted, we decided to load our provisions. Our food had arrived in Will Mason and Tim Thurnham's container together with their boat, *Bright Spark*. When we went to collect it, we could not believe the number of boxes. There was so much food that it looked like it would never fit. We carried the boxes back to *Yantu*, spread out a big piece of plastic on the pier and then started unpacking the boxes. We laid the food rations out in a matrix. Along the top we counted the weeks and down the side Monday to Sunday. Race rules required us to carry 90 days' worth of food. Each day consisted of a breakfast bag (containing either porridge or muesli, and, for Sundays, a tin of baked beans with sausage, bacon and bread), a lunch bag (containing crackers and pâté, pasta or couscous), a dinner bag (containing freeze-dried curry, Mexican chilli and the like with mashed potato powder or pasta), a night bag (containing powdered soup with crackers or instant noodles) and two day bags (containing sweets, pasta, biscuits, crisps, and isotonic powder to mix in our drinking water). It seemed like we would have no time left for rowing if we had to eat that much food each day!

As we started laying out the food on the sheet, it immediately became apparent that some of the food had suffered in the heat and melted together. There was not much we could do about that. We would simply have to munch the mixture! All this attracted the attention of four local boys who came over and sat next to the plastic sheet. They should not really have been there, but security on the pier was not as good as it should have been. One of them was clearly trouble and egged on one of the other younger boys to steal one of the bags. Luckily I saw it and we gave the boys hell. They eventually left, but it was stress that we could have done without. We packed the food into *Yantu*. Days 1-10 were placed in the fore cabin, then we moved on to the central fore storage compartment, which could hold about 20 days. Then the side compartments in the deck and the aft-central storage compartment where the emergency fresh water was also kept. It took us about four hours to fit it all in and at the end we still had about 10 days' worth of food left. We put this in black plastic bags in the sleeping cabin and then some more out in the fore cabin. There was food *everywhere!*

Tenerife — meeting the competition

(Left) There was food everywhere. (Right) 13 weeks of food lined up, ready to be put into the hold.

Pleased with our work we decided to go for a row. The crane complained as *Yantu* was hoisted over the side of the pier, but she went into the water without any mishap. We then got in and off we went. Boy, was she heavy! The addition of approximately 250 kg of food could certainly be felt at the oars.

We rowed back in and as we approached the pier, Rik Knoop, one of the Dutch rowers, yelled down to us that our water line was completely off. I jumped into the harbour, swam away from *Yantu*, turned and had a look. Rik had a point. We looked like a submarine about to do a very rapid nosedive. Something had to be done. *Yantu* was too heavy at the front.

We got *Yantu* back on land and suffered the usual, and now routine, damage to the skeg. Once repaired, we discussed the weight problem. There was no way we could take all the food and since we reckoned we would do the race in 56 days, we therefore decided to take out 15 days' worth of food. There was nothing in the race rules which prevented us from doing this, as they simply stated that we had to carry 'Sufficient food for at least 90 days'. Since the calorie intake was not specified we could thereby stretch our 75 days' worth of food to 90. That decided, we removed the food, mainly from the front, and re-launched to check the water line. We were still slightly front-heavy, but overall we looked a lot better. Exhausted we went back to the flat to crash. Bettina and Véronique cooked us an excellent dinner with the excess rations.

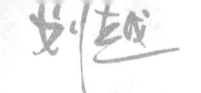

About to suffer damage to the skeg, as Yantu is put back on her cradle.

Three days before the race start it was time for scrutineering. This involved staff from the Challenge Business coming around and checking that all boats were seaworthy and equipped according to the rules. Matthew from the Challenge Business inspected our boat. Except for having to buy extra light bulbs for our torches and putting a wooden crate under our life-raft, we passed without trouble. He even commented that our water-maker installation was the best he had seen! Having our hard work endorsed felt good. As he was leaving Matthew asked: 'How about your EPIRB. Is it OK?'

'Sure, we tested it in Hong Kong and it is fine, but we can test it again if you like,' I replied. I threw the switch to the test position and awaited the green flashes. Nothing happened! Shit! At about US$1,500 the EPIRB is an expensive piece of equipment. It is also very key to safety as it is capable of emitting an automatic and ongoing SOS signal. There was no way we would have time to get a new one.

'I'll take it with me and get it serviced,' Matthew volunteered. 'In the meantime you can have one of our spares for the duration of the race.' What a relief!

Tenerife — meeting the competition

Matthew left and an elderly gentleman came over. He introduced himself as an expert on solar panels and told us he was helping out the Challenge Business with the scrutineering.

'How much power do you have in your solar panels?' he asked.

'We have two 30 Watt panels,' I replied.

'You will never make it across with that. The Inmarsat D+ draws more current than originally estimated. You simply do not have enough power,' he replied in a concerned voice and proceeded to lay out his calculations, which, given I am no electrician, I did not understand. But it sounded right and it scared the living daylights out of Sun Haibin, Véronique and me. What to do?

I called Duthie in Hong Kong, who had put the electrical system together: 'Duthie, they say we do not have enough power to run the Inmarsat D+. Are you sure you got the loads right?' I asked in a voice close to panic.

'If the Inmarsat D+ draws more than what we were originally told you may be a bit low, but you don't need it on all the time. Switch it off when you are not being polled and you will be OK,' Duthie replied. He sounded a bit annoyed, almost as if he thought we were going to use this as an excuse to cop out of the race. I thanked him and hung up.

We walked down the line of boats and asked the different competitors how much power they had from their solar panels. It confirmed our worst fears. By some margin we had the least of all! Shit, shit, shit! We were now really concerned. We had to get more power!

The next day Véronique, Sun Haibin and I drove all over Tenerife looking for solar panels. We found plenty, but none were for marine use. As the day progressed we became increasingly disillusioned and when we returned home late at night, and having exhausted the possibility of getting any flown out in time for the race start, the mood hit rock bottom.

'I just cannot believe Duthie got it wrong. It simply has to work,' I said, as there was little else we could do. Neither Sun Haibin or Véronique looked convinced and we discussed what to do. In the end we decided to buy 100 AA batteries so that we could use these for the GPS, torch and Walkman. That would only lessen the load a little, but it was the best we could do. We would have to watch our power very carefully!

Chatting with Graham Walters put our problem into perspective. Graham had successfully completed the 1997 race and therefore commanded a certain amount of respect. In 1997 he had arrived in Tenerife with an incomplete boat, the *George Geary*. To the disbelief of the other competitors he continued building his boat on the pier and somehow managed to get it ready in time for the race start. And he had actually finished tenth in the race!

Graham was now back on the pier with his *George Geary*, but it was in a terrible state.

'It looks like you left it in your back garden and forgot about it after the last race,' I commented.

'Well, actually I did. I was not planning on rowing it again, but then Michael Ryan contacted me and told me his partner had pulled out. So we made a deal. I would throw in the boat, he would pay the race fees and we would then prepare the boat here in Tenerife,' Graham replied.

I looked at Graham's boat. It was in bad shape. Bad paint job, no hatches, no solar panels, no nothing!

'Do you really think you will be ready to go?' I asked.

'Sure, I'll finish the painting today or tomorrow. Maybe I'll also find some solar panels to put on. That would be nice, but if not we can always operate the water-maker manually,' Graham replied, looking and sounding completely unstressed by the whole situation.

He was phlegmatic: 'After I finished rowing the Atlantic in 1997 I decided to learn how to row, so I joined a rowing club. It is a bit bizarre, really, to row across the Atlantic and then join a rowing club to learn how to row. My technique is definitely better now.'

I watched Graham over the next days. I never once saw him lose his good mood or seem the slightest concerned about his state of preparation. I think that to the end of my days I will always aspire to be able to deal with stress like Graham.

Eventually, Graham and Michael would finish 22nd in the race, taking 77 days to complete the crossing.

UPDATE

> Graham has since rowed the Atlantic four more times. In 2020, at the age of 72, Graham rowed 'George Geary' single handed across the Atlantic from Gran Canaria to Antigua and thereby became the oldest person in the world to row the Atlantic solo. What an amazing guy!

Tenerife — meeting the competition

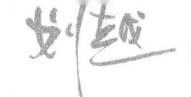

Thor Heyerdahl had retired to Tenerife so I hoped that we would meet up now that we had proven we were not fakes, and had actually raised a scholarship and made it to the race start. I sent him a fax asking if we could meet. I could see from Thor's address that he was living only about half-an-hour's drive away and felt sure he would come by or invite us round. But later that day I received a call from his housekeeper: 'Thor is busy writing his latest book. He has no time to see you.' What a disappointment! I later learned that he was terminally ill at that time (he died on 18th April 2002). That put things more into perspective. My own father was also terminally ill and also writing a book before he died, and the task of completing the book before it was too late became all-consuming. However, about to embark on the race, I did not know this and I felt bitterly disappointed.

The atmosphere on the pier was becoming increasingly tense as the race start drew nearer and tempers were flaring. It seemed that everyone needed a pet hate to let off steam about. For most teams this meant the Challenge Business, for others the Ocean Rowing Society.

The brothers Steve and Mick, rowing *Mrs D*, hated the Challenge Business and perhaps not without reason. They had used the recommended boat builder, who for some reason had removed the skeg from their boat. They had arrived early in Tenerife to prepare for the race. *Mrs D* was in mint condition and they had been feeling on top of the world until other boats started to arrive and they realised their skeg was missing. When they had discussed the issue with the Challenge Business the reply had been clear — get a skeg or you won't be allowed in the race. So from having one of the best prepared boats they suddenly found themselves out of the race before it had even started. Luckily, a British retired boat builder, Roger Whitehouse, happened to be walking down the pier one day and quickly got roped in, but it still took almost two weeks for the skeg to be added. In the meantime Mick and Steve partied hard to keep up their spirits.

UPDATE

> After the race Mick unsuccessfully tried to row the Pacific solo in 2003 and 2004. He then successfully rowed the Atlantic in a pair in 2005 and in 2009 he also successfully rowed the Pacific from Japan to California with Chris Martin. He then rowed with Steve Sparkes, a blind Royal Marine, from California to Hawaii in 2018. Apparently,

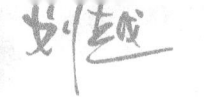

Steve is the first rowing partner not to complain about Mick's looks. Mick wrote a book about his voyage, called "Rowing the Pacific", so if you think this book is bad, try his (Mick and I remain good mates).

Our dealings with the Challenge Business had been positive throughout and we had always found them very responsive and helpful. However, the issue of the race start venue was now bothering everyone. The Challenge Business was adamant that the race would start from Los Gigantes despite what the harbourmaster said. They just would not accept defeat and move the start to Playa San Juan. We rowers could see the writing on the wall. We would stay on the hard stand in Playa San Yuan until the race start.

The title sponsors, Ward Evans, threw a fully-catered party for the rowers. Chay Blyth responded with a US$25-per-head party with a cash bar. That did not win him many friends, but then Chay was running a business. It was quite funny to observe Chay. Clearly he had achieved a lot and putting on the race had given us all focus and structure too. Having said that, all teams were paying top dollar for the privilege and we felt that Chay was making a thick margin. Chay seemed completely oblivious to this and would not understand why we were less than satisfied with how the Challenge Business was handling things locally.

The day before the race we had our final race meeting. Chay finally conceded that we would not be allowed to dock in Los Gigantes harbour prior to the race or start from there. However, he was not going to let it go. On race day he wanted us to row down past Los Gigantes, which meant a detour of four to five hours. That seemed plain stupid. We would be rowing along the coast to a harbour where we would be closer to La Gomera, an island west of Tenerife. La Gomera has the strongest currents of all the Canary Islands and rowing closer to it when there was no need seemed like a bad idea. The currents could trap us from leaving the islands. We rebelled. Chay countered that it was to please the sponsors, who had all their branding in Los Gigantes and none in Playa San Juan. The sponsors then stood up and said they would go with whatever decision the rowers wanted. Wild cheering followed. But Chay was not going to give in.

'I have done many races and I know you will look back on this race and be disappointed if you do not start from Los Gigantes. That is my

Tenerife — meeting the competition

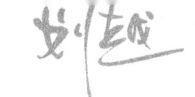

final decision.' Chay then drew a sketch map on the whiteboard of what the start would look like.

Before leaving he finished his briefing: 'The race will start at 10AM tomorrow. You will not have time to put your boats in the water then, so you must put them in today. We do not have any mooring in Playa San Juan, so you will have to take fishermen's buoys. This is a very rude thing to do and we cannot guarantee that they won't cut your boats loose.'

And on that note Chay finished his briefing. Not exactly a strong display of problem ownership. We were all his customers and he was treating us like bad employees. We looked at each other in disbelief, shuffled out the door, and went back to our final preparations.

Yantu was basically ready and we decided to avoid the queue that would certainly be building at the only available crane and put her in the water straight away. We then rowed her out and tied her onto the side of a fishing boat. The wind had been building over the past two days and I reckoned that no fishing boats were going to head out, so tying onto a fishing boat would be a safer bet than using an empty buoy as boats may come in from the ocean.

Having secured *Yantu* we swam back to shore. There I made a trip around all the boats with my needle-nose pliers. I had discovered in Hong Kong, when we had taken the water-maker apart, that only a certain diameter of pliers could be used to take out seals from inside the water-maker. It would be a real disappointment to have to leave the race because of not being able to repair a water-maker due to the pliers being too big, so I talked to all the competitors explaining the problem to them. Quite a few hurried off to buy new pliers. I felt good having done this. 'Surely this must earn us some good luck,' I said, looking towards the sky.

Véronique suggested we should leave the pier and go for a picnic inland to relax and forget about the race for a while. We all piled into the car and drove off. Halfway up Mount Teide we stopped, found a nice place under some pine trees, opened a bottle of wine and dug into the food. We could see the ocean stretching out to the horizon and beyond. We ate and fooled around for a while before heading back to the pier. It was dark now, but there was still hectic activity and probably only half the boats were in the water. *Yantu* was still securely tied up. We could clearly see her Chinese and Danish flags tied onto her radar reflector in the dusk. From the flags we could also see that the wind was continuing to build.

Relaxing and final carbo-loading on Mount Teide the day before the race together with Bettina (with cap) and Véronique.

We then concentrated on finding the spare EPIRB the Challenge Business was supposed to lend us. Eventually we got it and walked home to have dinner. The last thing Sun Haibin and I did before going to bed was to prepare our 'grab bag' — a waterproof bag, which contains personal documents, some supplies, a spare GPS and other useful things. It is called a grab bag because you grab it if you have to abandon ship and it is placed so that you can take it with you in a hurry.

Sun Haibin and I preparing our 'grab bag' the night before the race. The seriousness of the upcoming voyage is starting to sink in.

Earlier in the day we had also been reading the local English paper, the *Sun*. The front-page article reported that ocean rower Nenad Belic was feared lost at sea in his attempt to row solo across the Atlantic from America to Europe. This was a time for serious reflection about what we were about to embark on.

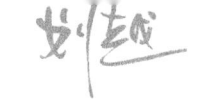

UPDATE

Since Nenad Belic, three more ocean rowers have been lost at sea. In February 2016 South African Mike Johnson, rowing Trans-Atlantic in a team of eight, was swept overboard by a large wave in high winds during the night. His safety ankle strap broke and he got detached from the rowing boat *Toby Walles*. He was not wearing a lifejacket. Due to the weather conditions the remaining crew were unable to turn the boat around and recover him.

In April 2020 the remains of Mainland Chinese Yu Ruihan attached by rope to his rowing boat washed up on an island in the Philippines. Ruihan seems to have been a guy who lived for the day. In 2015 he purchased his first ocean rowing boat in Greece with the intent of rowing across the Atlantic. He shipwrecked in Egypt. Undeterred, in 2017, he set out alone in a four people rowing boat from Sausalito, California, with the aim of rowing to China. 53 nautical miles off Maui he was rescued by a US Coast Guard helicopter. His boat was lost. Undeterred, he bought another boat in Hawaii, and set off again, only to be rescued again, having collided with a whale. His boat was lost, again! In 2019 he set off again from Hawaii in his fourth boat and capsized somewhere in the Pacific Ocean. The boat did not self-right and clinging onto its side he called the Hawaiian Coast Guard by satellite phone requesting rescue. He had lost his lifejacket and had not brought a liferaft. The Coast Guard found him and dropped a liferaft at the very limit of their flying range, so they could not stay with him. When they returned after refuelling they only found the liferaft. Ruihan's phone had run out of battery by then. Him and his boat were only seen again when they washed up on the shore in the Philippines five months later. By the time Ruihan died he was the Mainland Chinese ocean rower with the most days and miles at sea. In June 2020 American Angela Madsen's body was recovered by cargo vessel Polynesia, which had been directed to her boat by the US Coast Guard. Angela was sixty days into her solo row from California to Hawaii. When found, she was attached to her boat dead in the water. We know she had gone into the water to fix her bow drift anchor, but it is unknown why she did not manage to get back on board. Angela was an ocean rowing legend. She had already rowed the Atlantic Ocean twice, the Indian Ocean once, the Pacific Ocean once in a pair, and was into her second attempt at rowing the Pacific Ocean solo when she died. But what makes her truly special is that she did this despite being paraplegic due to a botched back injury surgery. She was also a Paralympian. On land she was in a wheelchair and at sea in her rowboat she was free. Her full story can be found at www.rowoflife.org.

Ocean rowing is a dangerous sport and if your time is up, that's it. Carpe diem and RIP.

After dinner we went down to the pier to do a final check on *Yantu*. She looked fine. We then stopped over at the American team's place, who were still busy sorting through stuff and discussing what to take and what not to take. We felt good that this was all behind us! We strolled back home and went to bed.

I did not sleep well, not so much because I was preoccupied with the race start. The wind was building and I was concerned that *Yantu* might be smashing into the side of the fishing boat and get damaged. Eventually I fell into a fitful sleep.

Sun Haibin showing the Chinese flag. He was immensely proud to represent China in the race.

The race starts!

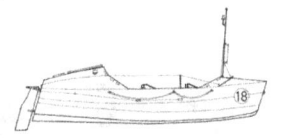

On race day we got up before first light. The girls cooked us breakfast, which we ate in silence. At about 8AM, I wandered down to the pier. It was a strange feeling to be heading off across the Atlantic and all I was carrying was a grab bag. My mind kept searching for critical items we had forgotten to bring, but I could not think of any. We were well prepared.

When I arrived on the pier there was chaos. Quite a lot of the boats were not yet in the water and rowers and supporters were rushing around trying to get ready in time. I looked out at *Yantu*. She looked fine. No signs of damage from where I stood. I also noticed that the flags were flapping less fiercely, a good sign that the wind had decreased.

Matthew from the Challenge Business came over to have a chat.

'The forecast says strong north-easterly winds. We have therefore changed the starting line. We will be starting off the Playa San Juan breakwater and head directly to Barbados from there. We won't start from Los Gigantes. With this wind it is too risky. You would get blown onto La Gomera and might not be able to get out of the currents there. We still start the race at 10AM local time.'

This was good news and made a lot of sense. It was now 8:30 and I started looking around for Véronique, Bettina and Sun Haibin, who had told me they would be right behind me. Véronique and Bettina arrived, but not Sun Haibin. Véronique told me he was on the phone to his girlfriend Cao Xinxin and would be down soon. We stood around waiting without speaking to each other. It was quite difficult to find something to say. About 9AM Sun Haibin finally arrived. He was looking preoccupied, but soon found his good mood again.

Some journalists came around to take pictures and did a brief interview, but we did not really register them. We were now trying to hail the tender, which would ferry us out to *Yantu*, but it was of course very busy. As the tender arrived Chay and the sponsors came down the stairs. 'I'll take that,' he said. 'You take the next one.' I just about blew a fuse. We had a race to start in, and he would be drinking beer watching us row. Perhaps he should give the competitors right of way! Matthew from the Challenge Business spotted the look on my face and told Chay the tender was big enough for all of us.

About to step into the tender and go to our boat. The next opportunity to step on solid ground is 5,000km away, hence the slightly preoccupied expressions.

Sun Haibin and I took off our shoes and gave them to Véronique. We wouldn't need them on *Yantu* and they were extra weight. I kissed Véronique goodbye and then Sun Haibin and I got into the tender. The tender was deep in the water, but delivered us safely before it headed out to the safety vessel with Chay and the sponsors. We climbed on board *Yantu* and were pleased to confirm she had not suffered any damage during the night. We put the grab bag into the cabin and prepared ourselves. There was not much that needed doing. We undid the moorings and untied the fenders which had held us off the fishing boat. Since we did not need the weight of the fenders we put them into the fishing boat as a thank you for having provided mooring for us for the night.

Before rowing out of the harbour we pulled out our secret weapons — funny hats. On a trip to Copenhagen, I had managed to find a horned Viking helmet and in Hong Kong we had purchased a Chinese coolie hat for the race start, to add some fun to a tense day with a subtext of promoting international understanding.

The race starts!

Hats on, we started to slowly row out of the harbour. We were in good time so there was no need to hurry. Some competitors were in more of a rush with their boats still on the pier.

Rowing out of the harbour to the race starting line.

A flotilla was building up at the starting line. It was quite a sight to see that many boats together. During the past weeks only a few boats at the time had been on the water. Now the whole fleet of 36 boats was gathering for the start and the sight was impressive.

Sun Haibin and I decided to hang back from the crowded starting line in order to avoid collisions and potential breakage of equipment. Losing an oar this early on would be bad news, even if we did carry two spare ones.

I checked the weather, which looked fine. I went over *Yantu* and she looked fine. I turned to Sun Haibin and he looked fine. Suddenly I was overcome with emotion. We had done it! Against all the God damn odds we were now actually here at the starting line, about to row across the Atlantic! The past one-and-a-half years had been such a huge effort and uphill struggle and we were finally about to reap our reward. I turned to Sun Haibin and shook his hand. 'I am so happy to be rowing across the Atlantic with you. I cannot imagine doing it with anyone else. We are a good team and we are going to succeed,' I said with tears in my eyes.

Sun Haibin was taken aback by this display. He thought I was sad about leaving Véronique behind and said: 'Don't be so sad about leaving. I was just on the phone to Cao Xinxin and she was very sad too. That was why I was late. I told her we would be OK and that I would see her soon.'

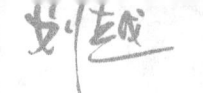

I tried to explain to Sun Haibin that I was joyful because we had managed to get it all together. It would take a while before he understood what I meant.

I just could not wait to start. When that gun went off, we would enter a world of our own. There would be no one to stop us or make things difficult. No lawyers, no extra invoices, no well-meaning advice to worry us sick. A simple, fair and apolitical world was ahead. Basically, all problems until Barbados would become binary. Would we get on? If no, could we work it out, or would we need to give up? Would our bodies break? If yes, could we repair them and continue or would we need to give up? Would the boat or equipment break? If yes, could we repair it or would we need to give up? Simple, simple, simple. Roll on the start!

The Challenge Business came on the VHF saying the start was delayed until 11AM because some of the boats were not yet at the start line. This was disappointing. A race is a race and if you are not there when it starts, then that is your bad luck. The Challenge Business received a lot of abuse back over the VHF from a number of teams. We sat back preparing to wait for another hour. The nail on my right big toe was looking long, so I picked it. It started bleeding, but it did not matter. Then the Challenge Business changed its mind again. We were now going to start at 10:15AM. We got busy.

I looked up to the starting line and reckoned it was a two-minute row to get there. We decided to row up from behind and pass the start line at full speed just as the gun would sound. The time drew nearer and the adrenalin starting pumping. Two minutes! We started rowing, but then out of nowhere the South African entry appeared right in front of us. We had to slow down to avoid colliding with them. This was not in the plan and we both felt frustration. When the gun sounded we were still about one minute from the starting line. Shit! We felt bad, but at least all risk of collision at the start line had passed by the time we got there, since all the other boats had left. We were one of the last boats to get across.

The race starts!

Rationally, it was a 50 to 60 day race, so one minute could not mean the difference between winning or losing, but emotionally it felt like that. We rowed on in silence, determined to catch up some of the boats.

A tender came up and followed us. It was Rob Hamill. About a week before, Rob and his wife Rachel had gone for dinner in Los Gigantes. As they were walking home Rob spotted a husband and wife violently fighting in the street and he interfered. The husband then attacked Rob, who hit back and broke his hand. It had been touch and go whether he would recover sufficiently to be able to do the race and the Kiwis had flown in a replacement rower, Mat Goodman. Even the night before the race, none of us knew if Rob would be rowing. We felt bad for Rob when we saw him in the tender. He must have been completely bombed out not to be able to row, having been the driving force in putting the two Kiwi entries together.

Rob yelled across to us: 'You guys look great. Go for it!' We waved back and pulled on the oars with renewed force.

Rob Hamill (wearing a hat) seeing us off at the start of the race. He had to abandon the race at the last minute due to a broken hand. Notice how the boats in the background are heading due West as we head South.

Next, Duthie's prediction that something would happen on race day to make our routing plan go to shit came true. Because of the north-easterly wind the fleet was heading almost due west. Our plan was to head south to below 21 degrees and then west, running with the wind. It was very discouraging to see everyone head west as we were going south. Were we doing the right thing? Should we head west, too? I remembered a golden rule from when I did my Yachtmaster certificate — *Plan your sail and sail your plan.* This must be true for rowing as well, I reasoned. There must be a 'Plan your row and row your plan' rule for ocean rowing. We continued south, but the nagging doubt stayed with us for the next two weeks.

The Dutch team was also rowing south and we were slowly gaining on them. This made us feel good. Suddenly, we gained a lot on them. It looked like they had stopped rowing and were heading back. Strange. We rowed past them at some distance so we were unable to find out what their problem was. We then spotted *Troika Transatlantic*. We slowly gained on them, but as we drew level they changed course and headed west. We were now leading the quest to go south and there did not appear to be anyone following us. Were we doing the right thing?

A hydrofoil full of supporters was approaching fast. It had come to see us off. I spotted Véronique on the top deck and waved at her. She waved back in between taking photos. I felt funny in my stomach. I was going to miss her. The hydrofoil stayed with us for a few minutes and then it went west to look for the other boats.

We watched the hydrofoil speed off. Sitting on our rowing stations our line of sight was about one metre above the surface of the sea. This meant that we could not see very far. Sometimes a wave would lift us up and we could see the other boats, but about six hours into the race we lost all visual contact with them. The next time we saw a rowing boat was 56 days later. We were alone in a 1,000 kg, seven-metre long and 1.8 metre wide rowing boat, with 4,000 metres of water under the keel and with 2,700 nautical miles to go until Barbados. We could still see Mount Teide on Tenerife sticking up over the aft cabin as we were rowing away from it, but even that was getting smaller. Increasingly all we could see was water and our journey was just beginning....

The race starts!

We soon lost visual contact with the other boats. Rowing away from land knowing the next opportunity to step on land was 5,000km away was quite unnerving. On top of that we were heading off in a different direction from the rest of the fleet. Some mental challenge ...

A stormy beginning

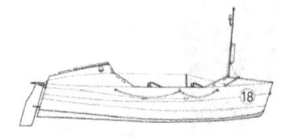

As we got further away from Tenerife, we came out of the wind shadow of Mount Teide and the seas began to build. It was now close to 7PM and we had been rowing together for almost eight hours. We decided to start rowing shifts. I dropped out to prepare dinner. From the side locker I took out a food bag marked WEEK 1, DAY 1 — DINNER. It was sweet and sour chicken with boiled rice. I poured it into the pressure cooker and added water, lit the stove, sat down on the side locker to watch the cooking.

'Christian, I can't row if you sit on the side. It makes the boat heel over too much, I can't clear the oar out of the water,' Sun Haibin said almost immediately.

I looked up. He was right. We were heeling 10-20 degrees to the side where I was sitting. I stood up and moved into the centre of the boat. I was now standing in the well, feet apart to keep my balance, supporting my back against the hatch into the aft sleeping cabin. *Yantu* straightened up and Sun Haibin was able to row again.

I took stock of the experience. Despite *Yantu* weighing about 1,000 kg, she was still very tippy and if I moved it had an immediate effect on the water line, which then in turn affected Sun Haibin's ability to row.

'It looks like we will have to do all our cooking standing up like this,' I said.

'At least it is mostly instant food, so it won't require much time to cook,' Sun Haibin replied.

That was the start of our four-times-per-day-stand-up cooking routine. Feet wide apart against the sides of the well and back firmly against the cabin hatch to steady ourselves. From this position we could then watch the gimballed stove swinging right next to us to ensure the fire did not go out and also stir the food. Because of the high aft cabin, it was a quite sheltered and comfortable position and since it was facing forward, the cook could have eye contact with the rower who was facing backwards.

(Half) naked chef Sun Haibin showing off our cooking equipment.

'OK, I think the food is about ready. Stop rowing and let's eat,' I said.

'Is it OK to stop rowing? I mean, the waves are getting bigger,' Sun Haibin replied.

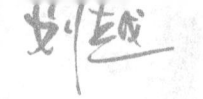

I had been preoccupied with the cooking and since the waves were coming up on us from behind, I had not really been noticing them. I now turned round and looked back. He had a point. The waves had certainly got bigger, but they were not white on the top so there was no danger.

'Sure, those waves are fine. Let's eat,' I replied.

Sun Haibin stopped rowing, pulled in the oars across the boat and tied them together with a shock cord so that they could not slip back into the water. He then came across and sat down on the side locker opposite to me in the well. I had taken out our dishes for the trip, which consisted of two cylindrical Tupperware plastic containers, chosen to prevent food from spilling out when *Yantu* was rolling from side to side. Sun Haibin held out the containers and I ditched food into them. We then settled back against the cabin and began eating.

This was the first real rest we had had in eight hours so it should have been enjoyable, but it was not. Now that we were relaxing, we began to realise that we were feeling seasick. Eating made us feel even worse, but we both knew we needed the energy and with much difficulty we finished the meal.

While we were eating, *Yantu* had turned sideways to the waves and the sight was getting impressive. Big dark walls of water were constantly approaching. As a wave got so close that you would have to strain your neck backwards to see its top, *Yantu* would start rising like an elevator and pass neatly over the top. Another elevator ride down into the trough at the back of the wave and then the next wall of water approached. Up and down. Up and down. I studied the waves for a while. Some of them now had white caps on the top, a sign that the wind was building and that things would get increasingly uncomfortable. It was also getting dark.

In Tenerife we had discarded the idea of wearing a life-jacket at all times. It was simply not practical to row with. The oars got caught in the life-jacket, it was hot to wear, and it increased friction against the skin, something I will come back to later! However, that did not mean that we weren't serious about safety and we had agreed on three safety levels when rowing, depending on the conditions. Level 1 was for reasonable weather and daylight. The safety factor comprised of being protected by *Yantu*'s hull and the wire railing. Level 1 was automatically escalated to Level 2 once it got dark. Level 2 safety consisted of tying ourselves onto *Yantu* with a rope around

A stormy beginning

the ankle and not being allowed to go to the toilet without first making sure the partner in the cabin was awake and alert. If one of us fell overboard at night, it would be unlikely that he would ever be found again unless it was noticed immediately. Quite a few sailors who drown have had their bodies recovered with the fly open or pants down. They go on deck to take care of business without telling anyone, fall in and are history. With a boat like *Yantu*, the risk associated with falling overboard was even more serious. Because the highest point on *Yantu* was only about 1.5 metres above the surface, the person in the boat would not be able to spot anyone in the water very far away, and the one in the water would quickly lose sight of *Yantu*, too. In addition, if it was blowing more than a force 4, the person left in the boat would not be able to row back against the wind to recover the man in the water. Being tied on by the ankle could not prevent us from falling overboard, but it would prevent us from becoming separated from the boat. We could then either yell and scream until our partner came out of the cabin to help or, weather permitting, climb back on board by ourselves. Level 3 safety consisted of wearing a life-jacket and being tied to the boat by a lifeline from the life-jacket to the jackstay. This level was used day and night when the waves were big and starting to break, as this presented a threat that we might be swept out of the boat or capsized.

I took another look at waves, felt the wind on my face, and reached into the sleeping cabin to pull out my life-jacket and lifeline. This was Level 3 conditions.

VIKING had given us great life-jackets. They were similar in design to those used on airplanes, so the bulk of them did not interfere that much with our ability to row. They had another smart feature. Typically, the inflation activation mechanism for a life-jacket consists of a small calcium pellet that dissolves when wet, which allows air from the gas canister to enter the life-jacket and inflate it. The down side to this is that if you get hit by a wave or if it is pouring with rain, then the pellet also dissolves and the life-jacket inflates, which is a bit of a nuisance. Both things were likely to happen to us often so the dissolving pellet would therefore be of little use. Instead, we had an inflation system which responded to pressure. Only if we got one foot under water would the life-jacket inflate.

I put on my life-jacket, sat down on my rowing station, tied on the ankle strap and clicked the lifeline on to the jackstay. I was ready to go.

I started rowing on the left oar to turn *Yantu* in the right direction. Because of the high cabin at the back catching the wind this was hard work. *Yantu* would much rather sit sideways on the waves than run with them. Eventually I got her around and I started rowing on both oars.

While I was turning *Yantu*, Sun Haibin had been busy doing the dishes. Dishes were cleaned with salt water and rinsed over the side. The waves were too much for Sun Haibin and he had already thrown up his dinner. Looking miserable with seasickness, he went into the cabin to rest.

It was now almost completely dark, but thanks to the stars I could still make out the approaching waves against the sky. And they were getting bigger. I started to feel seasick and threw up. The only good thing at this point in time was that I did not have to wash my puke away, because waves were washing onto the deck taking care of this for me. I knew I could not afford to lose the energy the food had contained, so I reached into one of the day bags and started eating some sweets. I threw up again. Another wave arrived and washed the puke away. I ate some more and the cycle repeated. I was starting to feel really bad and it was also starting to get cold, but I kept rowing.

The first three weeks are the hardest, I kept telling myself. When I had done statistics on the previous ocean rows, I had discovered that everyone who gave up did so within the first three weeks. It made sense. If you cannot make the adjustment from land to sea and get to terms with what you have embarked on, then you will know so within three weeks. If you can then, after three weeks, being at sea becomes everyday life and you are therefore able to go the distance. It was just tough luck that we had to get such bad weather straight away. This was really testing our resolve.

The waves were big and dark against the starlit sky and starting to look frightening. Sometimes one of the big waves would light up white on top as the crest was pushed down its front by the wind. The waves were starting to break and that was a bad sign. *Yantu* would not be able to elegantly ride over the top of a breaking wave. Instead the wave would smash straight into her. More and more white flashes started to appear all around me. It would only be a matter of time before we would be hit.

When I saw it some distance away, I knew the time had come. This was going to be a direct hit. It came towering up over the aft cabin,

A stormy beginning

Yantu started to accelerate down the front of the wave and then the top of the wave broke and came rushing at full speed at us.

'Grab on,' I managed to yell.

There was a big WUUUUUSH as the wave rushed towards us, followed by a loud BANG! as it smashed into the aft cabin. It then travelled over the cabin and hit me in the chest. For a moment I was sitting in water, not able to see *Yantu* below me. Then the wave was gone again and water quickly drained out of the boat.

'What was that?' Sun Haibin yelled from inside the cabin.

'Just a big wave. Nothing to worry about,' I yelled back. But I was not feeling as cool as I was sounding. That had been a big wave and we were sure to be in for more of the same.

I rowed the remainder of my shift and endured several more direct hits. Five minutes before watch change, I called to Sun Haibin to tell him to put on his gear, including a rain jacket. Some time later he appeared through the hatch, which by now was kept shut because of the waves. He looked around. Conditions had significantly worsened since he had gone in to rest two hours earlier.

He now looked at me and with a worried voice he asked: 'Should I be scared now?'

Sun Haibin had never been out of sight of land until this moment, so it was a valid question. I was not feeling too good myself, but I was the person with the blue-water experience so it was my job to keep up morale.

'No, this is pretty normal. It will pass. Don't worry,' I replied.

Sun Haibin climbed past my rowing station to occupy the forward station, clipped on, strapped his feet in, and started rowing. We rowed together for a little while and then I stopped, tied off the oars, unclipped from the jackstay and went to stand in the well to take off my life-jacket.

'Will you be OK rowing by yourself?' I asked.

'No problem,' Sun Haibin replied with a tense and determined face. I never asked him again.

I climbed into the cabin. The transformation was instant. Despite only being protected by six millimetres of plywood I suddenly felt quite secure. My next task was to get comfortable. This was difficult as *Yantu* was constantly rolling 40 to 50 degrees from side to side and to a lesser extent forwards and backwards, too. Immediately inside the hatch there was a little well to step into. I put my life-jacket in there and sat on the floor with my feet in the well. The cabin was about 120 cm wide at entry end narrowing into about 60 cm at the back. It was 200 cm long, with about 80 cm headroom. The roof's height also decreased towards the back to about 40 cm. In short, there was not a lot of space. I lay down with my head towards the back and my feet dangling into the well. It was not comfortable at all as I was being rolled from side to side. Finally I settled for the first aid recovery position, on my side with one knee bent, as this locked me into place. Lying down did not improve my seasickness and I hoped I would fall asleep soon.

Suddenly I heard the unmistakable WUUUUUSH sound of a braking wave. I heard Sun Haibin yelling. Then BANG! The wave exploded onto the cabin. I was picked up off the floor and thrown against the side of the cabin. That hurt. No wonder Sun Haibin had not looked great when he came out. Outside you could see the wave and ready yourself. Inside there was nothing to hold on to and you could not prepare for the blow. When the next wave hit I was thrown against the wall again. After two hours I felt like I had been put in a sack, strung from a tree and beaten by baseball bats. Although I was incredibly tired and feeling seasick, I was more than ready to go outside and row again! I put my head outside the cabin. It was cold and wet.

'How are you doing?' I asked Sun Haibin.

'Not too bad, but every time I eat I am sick straight away. Is that normal?'

'Yes, it's the same for me. Don't worry about it, but make sure you keep eating and drinking or you will become dehydrated.'

We continued rowing shifts through the night. The weather was still bad, but it did not seem to be getting any worse, which was a good sign.

Feeling bruised and seasick, I was rowing the early morning shift the next day when suddenly I saw a big fin stick up into the air behind the aft cabin. At first I thought I was hallucinating, but then the fin moved right and I could see the rest of the animal. We had just missed colliding with a killer whale! It was now about 20 metres behind us swimming perpendicularly to the waves. For a split second it was lifted up by a large wave and beautifully displayed on the side of it. It was a big fellow, somewhat longer than *Yantu!* Despite being seasick, my adrenalin got flowing and I yelled to Sun Haibin that he should hurry out to have a look, but he was too seasick to find a near collision with a killer whale of any interest. The whale continued on

A stormy beginning

its course. I thought it looked about as fed up with the weather as we were and I am sure that it would have ploughed straight into us had we not passed slightly ahead of it.

Had we been a split second slower we would have colliding with a killer whale.

I settled back into rowing and Sun Haibin came out to make breakfast. We ate it all, threw up some more and changed shifts. The sun was now out and I decided to run the water-maker to replenish what we had used in the past 24 hours. I flipped the switch and the familiar humming sound from the piston movement started. A little while later it started to produce water. Fantastic, it was still working. Having done the dishes, I climbed inside the cabin to check out the draw on the battery. The voltmeter was reading 13 Volts. I switched off the water-maker. The needle jumped to 14.3 Volts. I switched the water-maker back on and the needle fell back to 13 Volts. This was tricky. If the battery was drained below 12.4 Volts we would not be able to recharge it from the solar panels any more. We would therefore have to watch the voltage very carefully when we used the water-maker and we would only be able to run it if we had direct sunshine on the solar panels, or we risked the battery being drained before the sun came back. Would this work out? We would have to run the water-maker five hours each day to replenish our water. Would we have so much sunshine each and every day? We could only hope!

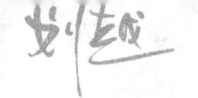

I explained the problem to Sun Haibin and instructed him to wake me up if cloud cover came along. I next turned to navigation. I spread the chart on the cabin floor. We had put it in a plastic pocket, because we knew we would not be able to keep it dry for the duration of the trip and we would simply mark our position on the plastic transparent cover with a felt tip pen. I turned on the GPS, read our current position and plotted it on the map. The plotting was difficult because of the wild pitching of *Yantu* and it made me feel very seasick, but the result was encouraging. In the past 24 hours we had covered 60 nautical miles! Had we not felt as bad as we did, we would probably have celebrated a lot more. Then a thought came back to haunt me. Were we right to be heading south when everyone else was heading west? I lay down and tried to concentrate on our choice of course, but I could not. After some fretful sleep and my quota of being smashed against the side of the cabin, I went back on the oars again.

The fast progress of 60 miles was fresh in my mind and, looking at the waves, I reckoned that with slightly more effort I would be able to get *Yantu* to surf down the waves. I tried it and it worked. When I caught a wave I was clocking seven and more knots on the log, compared to the average two to three. This was fun!

I made my first satellite call from the ocean to Teresa at the Challenge Business's race office. She told me the Spanish entry had been hit by a freak wave and had had to go back to Tenerife for repairs. The Kiwi girls had broken an oar. *Yantu* had not survived unscathed either. When I went back out on deck Sun Haibin showed me a broken stainless steel stanchion. We had tied a spare oar from the aft sleeping cabin to the stanchions on both sides of the boat in order to prevent us from falling out when we went forward to our rowing stations. The starboard stanchion and the oar were being relentlessly pounded by the waves and eventually the weld on the stanchion gave in. There were great forces at play!

A stormy beginning

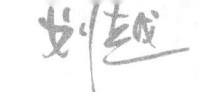

Great forces at play. Sun Haibin showing the steel stanchion that was broken by a wave.

Later in the day it started raining and it became cold. Sun Haibin was rowing and we agreed that I would cook us some pasta to get some warmth into our bodies. I took out a serving from one of the day bags and started to boil it. It was raining, cold and we were miserable. All we wanted was some good warm food to heat us up, but the damn pasta would not go soft. It took forever to boil it and once ready Sun Haibin stopped rowing and joined me in the well of the cockpit. We dished out the pasta and tucked in. Crunch! It was full of whole peppercorns and bay leaves. I spat it out and tried a new mouthful. Same thing. We tried to eat it, but had to spend most of the time spitting out peppercorns and bay leaves. The pasta was inedible! On the whole trip, Sun Haibin only lost his temper once and this was it.

'You have to tell Maggie about this pasta. This is totally unacceptable,' he said and threw the rest of his food overboard. I did the same. We looked at each other in disappointment, but neither of us had the energy to cook anything else. Later we cooked more of the pasta, but the result was always equally disappointing, and eventually we simply threw it overboard without bothering to cook it first.

On a later shift that day I had an out of body experience! I suddenly found myself up in the sky looking down on myself rowing. It was a dramatic view. I could see the ocean with the large white-capped waves stretching to the horizon. *Yantu* was only a small speck in the moving sea of blue and white. I watched the waves coming up from behind *Yantu*, tower over her, and then either slip gracefully under her or explode onto her. When a wave hit her both myself and *Yantu* would momentarily be lost from view in the hissing water, but then we would reappear and I would row on. Sometimes after catching a surfing wave *Yantu* would end up sideways to the waves and I watched myself pulling the oars with great effort to get back on course before the next wave hit us. As I was watching the scene below I started to feel increasingly calm. It was hard work and the elements were at their worst, but I could see that I was doing well in spite of this. It gave me confidence that we would weather the storm successfully. Feeling thankful for this insight, I returned to my body. I have no idea how the vision happened or how long it lasted. I have never before had an experience like that and never since. Maybe greater forces felt we were in need of some perspective of what we were doing? In any case it was timely and provided not only comfort, but a memory that will stay with me for life.

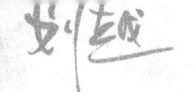

As night was approaching I was feeling increasingly tired. The lack of proper sleep, being seasick and expending the extra energy to get *Yantu* surfing was taking its toll. Eventually I was so tired that I simply could not get up to row. I had managed to row myself into the ground. *How stupid can I be?* I asked myself. I told Sun Haibin and since he was also feeling tired and the night was cold, we decided to put out the drift anchor and sleep until the morning.

Being two in the cabin was not at all comfortable. We constantly rolled onto each other and when the WUUUUUSH — BANG! waves hit us we would smash into each other as well as the cabin side. But somehow we managed to get some rest and the next morning we continued rowing. The consequence of our rest became apparent when I did the navigation. We had only covered 42 nautical miles on the second day. However, taken together, the two last days were on target, as we needed to cover 50 nautical miles per day to make it to Barbados by 3rd December.

The next day Sun Haibin caught on to the idea of getting *Yantu* to surf. As a result he also rowed himself into the ground. We simply could not keep up this way of rowing or our bodies would break for good. As we became more tired, we had less power to row and *Yantu* increasingly turned sideways to the waves, despite our best efforts to keep her on course. That night we decided to sleep again and the next morning, the result was immediately apparent when doing the navigation. We had only covered 30 nautical miles. Also, we had set the rudder wrong and had drifted towards the African coast. This caused my friends to wor-

A stormy beginning

ry when they logged onto the race website to check our race position. Véronique later told me that my friend Ole Bang had called her: 'Next time you speak to Christian tell him to watch his compass or to start learning Swahili!'

Suddenly I was looking down on myself rowing.

It was hell to get up the next morning. Everything was wet and we were feeling fragile from the constant pounding by the sea. We were very close to having spent all our energy and the weather was still not showing signs of improving.

'We need to change the way we row,' I said to Sun Haibin while he was rowing, as I took out the GPS to check our speed.

I then said to him 'Can't you row slower than that?'

He looked at me in disbelief: 'Give me a break!'

'No, I am serious. Try to pull more slowly,' I continued.

Sun Haibin slowed down. It made no difference to the speed.

'Come on, you can row slower than that!' I challenged him a second time.

Sun Haibin slowed down even further. There was still no effect on the speed, but he was now looking a lot more relaxed. This was working!

We spent the next several watch shifts challenging each other: 'Come on, tough guy! Can't you row slower than that?' Slowly we started to regain our energy. We did 37 nautical miles that day. Not enough, but more than the previous day.

I decided to call Véronique for the first time. It was great to hear her voice and it almost made me choke, so I had to hang up again as I did not want to worry her. She was still worried about the water-maker and whether we had sufficient power. She wrote in an e-mail exchange with my sailor friend Olav:

> *This morning I am a nervous wreck because I can see that their position has not been updated for a much longer time than the other boats. [We had actually been sleeping and missed switching on the D+ unit.]*
>
> *To be honest, I am not worried about their safety (not yet because they are not so far away from the shore) but my biggest fear is that they have technical problems with the electricity supply or the water-maker. If they had to give up because of that, it would be real, real drama for Christian. I am so scared of that!*

I fully believe Sun Haibin and I were in the easier position because we could see the waves and deal with the equipment problems. The shore crew could only imagine our problems and your imagination can easily get the better of you.

I took a seasickness tablet with a bit of hot sugar water and for the first time I managed to fall properly asleep. When I woke up an hour later it was as if a miracle had happened. I felt 100 percent fit and ready to go. On the whole trip I never got seasick again.

Another positive sign was that the wind was finally decreasing, which made rowing easier. However, Sun Haibin was starting to experience

A stormy beginning

back problems. We decided he would rest two shifts. He went into the cabin and went through the medical kit where he found some deep heat ointment. After applying the ointment he rolled a towel into a ball and put it under his back. The movement of the boat meant he was constantly moving over the ball, thereby being massaged. At least the pitching and rolling was good for something! A few days later Sun Haibin had recovered again and slowly he also got over his seasickness.

We finally could not take it anymore and one morning we decided to row west! I was rowing the early morning shift and now we were rowing west the waves were coming sideways onto the boat. It was still windy, so I was on a constant lookout for breaking waves. Suddenly I saw a wave with a white top, but it did not look quite right. As it came closer I suddenly realised that it was not breaking. The white cap was caused by eight whales swimming side by side. And we were on collision course! I stopped rowing, not knowing what else to do. About two metres from *Yantu* the whales suddenly decided to dive and with great grace they slipped under the boat and came up on the other side. They stopped swimming and turned around to have a look. For about 30 seconds I had eye contact with eight pilot whales, each the size of *Yantu*. We looked at each other in amazement and then I composed myself.

Eight pilot whales came bearing down on Yantu.

'Sun Haibin get out here now! There are whales right next to us. Bring the video camera,' I whispered as loudly as I dared. This time Sun Haibin was not seasick and he quickly came out with the camera. We were now further away from the whales, but we still managed to get some shots, before the whales and we continued on our respective journeys. What an experience. We were on a high for the rest of the day!

Despite the whale sighting it still didn't feel right to be going west so we went back to our southern course. We decided that we would not change from our original course again. We had to stick with one thing or the other, or we would drive ourselves crazy. *Plan the row and row the plan,* I kept telling myself.

That night the sea was finally flat, but we were rowing into some very nasty looking rain clouds. As we got closer the air became stuffy and moist. Two fish, which looked like blue eels, were jumping out of the water next to *Yantu*, leading us into the black cloud. When the sun set, we could no longer tell the sea and the sky apart because the black clouds closed in on the water. Thunder and lightning followed and the whole situation became increasingly eerie. We ate our supper and continued rowing in silence. Neither of us liked it. It started raining heavily, the temperature dropped dramatically, and the thunder and lightning got more intense. The air felt oppressive. Eventually we decided to take refuge in the cabin. We stayed in there from 8PM until 8AM the next morning. The weather had then cleared up again and this would signal the end of bad weather for a long time.

We were basically a mess for the first week of the race and it was very difficult to establish a rhythm. Véronique wrote in her supporter update about the first week:

If you have been watching the progress on the Internet, you already know that until now, they have always been within the top 15. Pretty good considering that most of the top 10 teams have a lot of rowing experience, unlike Christian and Sun Haibin.

They are getting on well together: 'better and better' were Christian's words today. Sun Haibin, who has no sea experience at all, is taking it all very well so far.

In the past week, they have been going through one challenge after the next: it was really interesting for me to observe how Christian mentioned a different type of problem with every phone call, and when I asked him about it 24 to 48 hours lat-

A stormy beginning

er, he had already forgotten that problem and was preoccupied by something different! Their 24-hour periods must seem much longer than ours....

First there was seasickness, which lasted a good four days and was very hard because it dehydrated and weakened them physically as they were not able to retain their food. Christian mentioned feeling so weak that he felt unable to move after lying down. Even making food was strenuous under those conditions. Then came the exhaustion from rowing too hard and wanting to surf the four to five-metre high waves too fast. Even though it was thrilling for both of them to be able to row faster after being sick, they quickly had to learn to pace themselves....

Subsequently came the physical pain from rowing so many hours a day, especially in the joints: knees, shoulders, elbows, hands. 'Just every joint in our bodies hurts so much, it is unbelievable'.

Then followed the difficulty of adjusting to the rowing/sleeping rhythm (approximately two hours of each activity, alternatively, so that there is always someone rowing the boat in the right direction and watching what is happening outside).

Finally there was the storm-test on Sunday, with rain, thunder, lightning and no visibility, which made it impossible to row for several hours. Both rowers took refuge inside the cabin. Although they were tossed around and had a hard time sleeping, Christian describes the cabin as a safe and dry place. He says that when you go in, it is similar to coming home to a fire when there is a snowstorm outside ... I wish this was really true.

Another day in the office.

Through the barrier

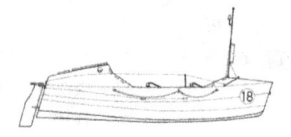

After the first week of rough weather, conditions had now turned almost ideal. A force 3-4 wind was blowing behind us, and three-to-four metre high waves rolled gently under us. Yet we were still not feeling good. The issue of our best course kept bothering us.

> *Log entry — Day 8 — 15th October*
> Wind force 5-6, later 3-4. Waves 3-4 metres. Barometer 1013. Cloudy with a little sun.
> Rowed all night and made good progress. Slept very well. Still not completely out of the warm front and the cabin remains stuffy. Would be good to get a clear sky and the ability to dry out the cabin. Luckily the solar panels work well through the cloud cover.
> PM: Am in a bit of a foul mood because of the navigation. Are we doing the right thing? Am listening to Spanish radio to cheer myself up on my off watch. Spoke to Véronique. She will soon make another update — she is so great and knows exactly how to support me.
> 18:00: Spoke to Olav and decided to head west. Rowing compass course 90°.
> Sun Haibin worried about navigation, but feels better now that he understands 'heading' is not the same as 'course over ground'.

Because the compass was mounted back to front in *Yantu*, when it was showing 90 degrees East, we were actually rowing 270 degrees West. This took some time to get used to.

Through the barrier

I was the navigator on board, but because we were feeling uneasy going south, Sun Haibin also started poring over the chart on his off-shift.

Log entry — Day 9 — 16th October
3:45AM: Changed course heading directly for Barbados.

Log entry — Day 10 — 17th October
Wind force 4, NNE. Waves 3-4 metres. Cloudy.
Hard rowing all night to try and keep on course. Very tired. Decided to change back to original course. Rowing West was a bad call. Plan the row and row the plan! We are now back on original course doing 2.4 knots. We are starting to feel good again.

Log entry — Day 11 — 18th October
Wind force 5-6, NNE. Waves 6-7 metres and breaking.
Tough night of rowing with lots of waves in the cockpit.

Log entry — Day 12 — 19th October
Wind force 4-5 NNE, decreasing. Waves 3-4 metres and messy. Sunny.
I felt low because I was tired and concerned about our course. Sun Haibin helped me get over it.

Log entry — Day 13 — 20th October
Wind force 3-4, NNE. Waves 3-4 metres and messy. Sunny.
Finally we got back on track. I feel a lot better.
16:00: Changed from 30° [i.e. 210°] to 0° [i.e. 180°] to get south ASAP and out of these messy waves.

Log entry — Day 14 — 21st October
Wind force 2-3 NNE. No waves. Cloudy.
No wind last night. Rowed myself into the ground and had to stop at 7AM, one hour early. Slept until 9AM — completely out.
Difficult to get speed into the boat.

Log entry — Day 15 — 22nd October
Wind force 2-3 NNE. Small waves. Barometer 1012. Sunny.
Finally got a bit of speed into the boat about 10PM last night. Comfortably doing 1.6 to 1.7 knots. This does not sound like much, but is a big improvement. Mentally we feel a lot better — and we broke through 21° North! Only one degree to go until the trade winds! I hope it will not be a disappointment. At this speed it will take us 70 days to finish. The Australians and Kiwis still believe they can beat the 41-day record, it's amazing that they can go so hard for so long. Really something to admire!

Log entry — Day 16 — 23rd October
Wind force 2-3 NNE. Waves small. Barometer 1014.
Good rowing last night. Finally the waves subsided and we could row 1.6 to 1.7 knots.
A bird landed on the boat at about 3AM, but left again. I think it got scared by the sound of the bilge pump when I pumped out the cockpit well.
Sun Haibin now has a pain in his right thigh — hopefully it will pass. I feel quite good.
Unexpectedly, I did not crash and burn last night. The rowing is definitely getting better as we are heading south.
Will call Véronique today. I did not call her during the last few days because I am afraid of what our position will be and also I was in a good rhythm that I did not want to break.

During the whole race, Véronique was writing (without my knowing) a diary of her experience of the race viewed from land, as a present that she would give me on arrival. The entry on 23rd October reads:

Really tough day again because of the race (there have not been that many so far, which is good!). I replied to millions of e-mails, the majority of which for Yantu.

The tough part of the day came from worrying — your position showed a very slow movement and a loss of positions in the race. Some horror scenarios started going through my head again — for the first time I started to wonder if you were still alive! Sounds silly, but I really had to control my mind not to panic. You have not called. I tried calling you. It did not work. MIND GAME! No need to panic, it just ruins my day. (I know this from experience now....) But still, I was quite nervous. Fortunately you called at lunchtime.

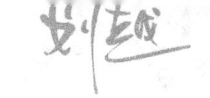

During the call I had to tell you that you had slipped back to 22nd position. Tough for you and I also felt bad about telling you that. It is hard to imagine what you are going through mentally and physically. Especially the lack of sleep must be awful. Whenever I feel exhausted, like tonight, I think of you and that it must be 1,000 times worse.

When I asked about physical problems you said your fingers hurt a lot. You can't stretch them or close your fist without it hurting. Sun Haibin is suffering from one leg. But you sounded like it was all bearable.

Apart from that, you said today that if you continue at this rate, you would not reach Barbados before 15th December.

After talking to Véronique I hung up and climbed outside to talk to Sun Haibin.

'We are in 22nd position. I just do not understand it. Of course we should lose places because we are going south and others are heading west, but down to 22nd! It just does not make sense.'

'Maybe we aren't as tough as we thought we were,' he replied.

'Come on. You saw the level of physical fitness of the other rowers in Tenerife. There is no reason why we should not be in the top 10. There must be something else. We used to easily be able to row two knots, but now we are struggling to row 1.5 knots! If feels like we are rowing in glue!'

'Maybe we have growth on the bottom of the boat,' he suggested.

I was dismissive of this as we had only been to sea for 16 days. I did not like the idea of stopping to check either, as that would slow us down even further. We ended up arguing, but quickly got over it and then we checked out the bottom.

We both looked over the side and the sight that met us was unbelievable. Under the water line *Yantu* was covered in an inch-long coat of sea grass and algae. How could we have got so much growth so quickly despite using antifouling? We quickly decided on two reasons. Firstly, we had seen the Kiwis lightly sandpaper the bottom of their boats on the day before the race. When we asked why they told us it was because it helped to activate the antifouling. We had thought it a bit excessive and had gone for our picnic instead. Now we were suffering the consequences. Secondly, the growth was probably extra strong because *Yantu* had been in Victoria Harbour prior to coming to Tenerife. Hong Kong's harbour is not exactly clean and a lot of restaurant waste finds its way into it. As a consequence *Yantu* had arrived in Tenerife with an — invisible to us — coating of cooking grease under its waterline. We could just imagine the party the plankton must have had as *Yantu* entered the water. 'Hey guys, free Chinese takeaway. Come and feed here!'

Having confirmed the growth, the next issue was how to get rid of it. We talked about it for some time and finally decided that one of us would stay on board looking out for sharks and holding on to a rope attached to the other person in the water. The person in the water, wearing goggles, would then scrape the hull below the water line with an empty tin can. Sun Haibin took the first shift. I tied a rope around his waist and he jumped into the water. Luckily the sea grass came off easily when scraped, so progress was fast. I stood up in *Yantu* looking all around. The sea was deep blue and calm. If a shark came from below I would have no chance of spotting it. From a practical point of view the safety rope seemed redundant. If Sun Haibin was going to be attacked by a shark I would not be able to yank him out of the water. But maybe the rope would enable me to recover a body part or two.... However, when we changed shifts and I got into the water to continue the cleaning, I found that though the rope might not have much practical use, the emotional support was high. When I came back up into the boat I joked with Sun Haibin about the rope, body parts and the sharks, but he flatly told me not to make jokes like that again, a request I subsequently respected.

Sun Haibin got back into the water to complete cleaning the skeg. This involved diving under *Yantu*. I had tried to do this, but my right ear did not feel comfortable, probably a left-over from the bends from my diving accident. Back on deck again, Sun Haibin volunteered to do the cleaning from then on. He liked the swim, but a more important factor was that he was allowed to use a litre of fresh water to wash afterwards, a luxury we had decided to otherwise only enjoy once a week in order to preserve drinking water since we were still uneasy with the power consumption of the water-maker. Sun Haibin found the infrequent washing with fresh water difficult to endure and the opportunity to get two washes per week was therefore very appealing. I also think he volunteered for another reason. I had put the whole project together and financed everything. I was the navigator, had the blue-water experience and maintained contact with the shore crew. Sun Haibin was looking for something unique which contributed to the overall success of our project that I could not do. Because of my ear problem, cleaning the hull seemed to be that, and

Through the barrier

from then on Sun Haibin cleaned the hull. Initially it was every two days, but as the scraping on the hull kick-started the antifouling, we gradually reduced the frequency to every five days.

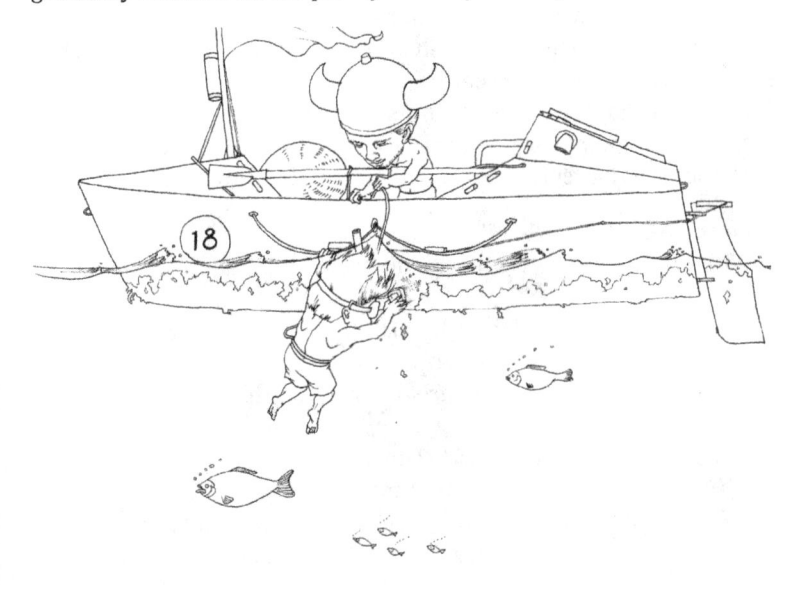

Cleaning the bottom of the boat.

After cleaning the bottom, we got into a home-improvement mood. I decided to tackle the scuppers. The scuppers consisted of two rectangular holes in the side of *Yantu* just above the water line. Their purpose was to ensure water ran out when we got swamped by a wave. However, when a medium-sized wave hit the side, water would also come gushing in. While Sun Haibin was washing, I took out a spare jerry can and cut out four rectangular pieces. I then took a screw driver, some screws and leaned over the side and screwed plastic flaps over the scuppers so that they were secured at the top and able to open outwards at the bottom. It took forever to do because of the movements of the boat, but eventually it was done. We now had a drier boat. From now on when a wave hit the outside of the boat, its force would slam the plastic flaps shut and no water came in. However, if a wave got into the cockpit the force of the water would push the plastic flap outwards and the boat would still be self-bailing.

The final improvement we made was to replace the broken stanchion on the starboard side with the undamaged one from the port side. The reason for this was that waves always came over the starboard side and it was therefore more exposed. In addition, as the hatch into the sleeping cabin opened up to port it was easier to walk forward on the starboard side after exiting the cabin. We were very pleased with the result, especially when we started rowing again. My log for the rest of the day reads:

Our speed is now at 2.4 to 2.5 knots — big improvement! We should get to the waypoint early tomorrow at this speed. GREAT!

The ocean no longer felt like glue and we were back to our original speed.

> *Log entry — Day 17 — 24th October*
> Wind force 3, NNE. Waves 1-2 metre and regular. Barometer 1014.
> Rowed 2.5 knots all night! GREAT! Got to waypoint N20° W25° at 4AM this morning. GREAT!
> The 17th day is over and we are doing better than ever!

> *Log entry — Day 18 — 25th October*
> Wind force 2-3, NNE. Waves 1 metre. Sunny.
> Rowed 60 nautical miles in the past 24 hours! Tired, but feels good to make such progress. Maybe we will make it to Barbados for Véronique's birthday.

A few days later, we passed the magic 21st day at sea. Statistically, we were now certain to now make it across! We took stock of what was happening in the race.

Three teams had given up and the crews taken off. A number of boats had gone back to Tenerife for repairs before restarting and others had been re-supplied with water or spare parts and were therefore also officially out of the race. It had surely been a tough first three weeks.

Dom, the other rower from Hong Kong, was one of those who had given up. He and his rowing partner and friend for 20-odd years, Jon, had not been able to resolve their different performance objectives and they had fallen out badly. So badly that Jon had asked Dom to leave the boat for one of the rescue vessels. Jon wanted to continue the race on his own and rowed on for several weeks, but eventually gave up and called the rescue vessel. Dom was on the first one that arrived, but Jon did not want to be on the same boat as Dom so he waited a while longer until the second rescue vessel came before getting off. Their boat, *Star Challenger*, was then burnt and sunk at

Through the barrier

sea as it was too far offshore to be towed back and it would be a hazard to shipping if it had been left floating around. I later met Jon in Barbados and asked him about the atmosphere on board.

'If we had had the space to fight, we would have. But since we didn't, all we could do was to sit in the cockpit and yell at each other,' Jon replied.

On 28th October, the *South China Morning Post* ran two separate articles on Dom's and our story on the same page. Dom's caption was **The pain and pride in calling it a day.** Our caption ran **milestone turns tide in favour of rowers' mammoth task.** It then went on to say:

> Despite the difficulties, the crew are still enjoying the journey — sort of. 'It's not like our sides are splitting from laughter, but we are having fun already. I guess it is all a question of how you define fun,' Havrehed said.
>
> The pair communicate in Putonghua, and their double motto, written in characters above the entrance to the cabin reads: 'Take it easy' and 'Work as a team and you will achieve more'.

The 28th October South China Morning Post article. English pair Jon and Dom's childhood friendship falls apart, as our unusual cross-cultural Sino-Danish partnership goes from strength to strength. Sometimes the most natural team is not the strongest team.

One day Véronique told me over the satellite phone that Andrew Veal from *Troika Transatlantic* had retired from the race because he could not shake his fear of the sea and confined spaces and that Debra was continuing by herself. Unbelievable!

'What a disobedient wife!' Sun Haibin said on hearing the news.

'What do you mean?' I replied, completely taken aback by his reaction.

'They are a married couple. By rowing on she makes her husband look weak. It is a double-handed race. You start and finish as a team.' Sun Haibin continued.

He had a point. It would not be easy for Andrew to go down the pub after this without being poked fun at.

'I think it is a great display of love by Andrew to swallow his pride and let Debra row on. It takes guts,' I observed.

However, as Debra continued her row even Sun Haibin could not help to admire the disobedient wife. Debra would complete the row single-handed in 111 days and through her determination and the press-concocted heart-rending story of her being 'abandoned at sea by her husband' she went on to row her way into the hearts of the British nation and subsequently on to fame and glory.

UPDATE

> Debra subsequently became popular on the speaker circuit and her marriage to Andrew did not last. It cannot have been easy for either of them to get up in the morning knowing that Debra yet again would be off to give a talk about how Andrew had abandoned her at sea and how she had succeeded without him. She published a book called "Rowing It Alone: One Woman's Extraordinary Transatlantic Adventure". Andrew's failure was her glory. I bet neither Debra nor Andrew had thought of this outcome when they signed up to do the race together, but it goes to show that if you decide to row an ocean, unpredictable things can happen to you in unexpected ways, because it will inadvertently rock the proverbial boat of everyday life.

Through the barrier

We felt sorry for the teams who had not made it past the 21-day barrier. An unbelievable amount of effort and commitment had gone into getting to the race start and that some teams were not going to make it across was tough to accept.

We took stock of what we had achieved so far. In the past three weeks, we had been tested to the limit. We had been seasick, hungry, cold, battered, exhausted, irritated, worried sick about our route, lost positions in the race, endured storms, and suffered damage to *Yantu*, but we had pulled through. As we passed the 21st day, we felt strong. The issue of whether to row south or west had solved itself as we were now so far south that we had no choice but to row west. We had overcome our seasickness, repaired *Yantu*, got speed back up and found a good rhythm. The first three weeks had been tough, but we were now a stronger team for it. As we turned the corner at the N20° W25° waypoint and headed west, we also turned over a new leaf — we were going to claw back positions!

We reached the waypoint and finally head west.

Heading west

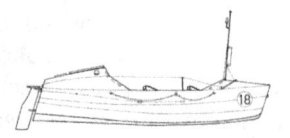

After we turned the corner at the first waypoint, we started making good progress towards Barbados. Most days, we were clocking up close to 50 nautical miles, almost enough to get us to Barbados in time for Véronique's birthday. As things became more routine, we started looking for ways to improve our performance.

> *Log entry — Day 21 — 28th October*
> Wind force 2-3 NNE. Waves 1-2 metre. Barometer 1012. Cloudy.
> Shortened the oars by 3 cm and I now have a better gearing. Sun Haibin complained that his rowing position is too high. Moved the life-raft to compensate.
> Saw a large fish — probably a marlin. Beautiful. My big toe has been bleeding and spilling pus since the race start. I should never have picked that nail! Must call Brian to get advice.

The cabin was very damp. The folded towel I was using as a pillow was so wet that I dreamt I was lying with my head in cow dung. When I woke up, I was not sure whether I was relieved or disappointed. Had I been lying in a field with my head in cow dung, it would have been embarrassing, but at least I could have gone home to shower and then to bed. Now I had to go outside and row. Sun Haibin laughed when I told him the story.

Heading west

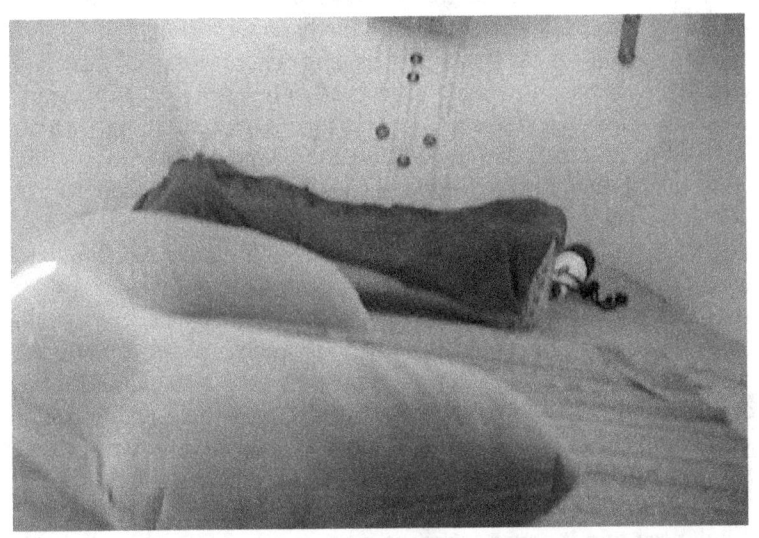

Our luxury bedroom. Seen at the back is a traditional Chinese pillow made out of bamboo, which was excellent because did not get wet.

Our Croker oars consisted of a long blade section with a shorter handle, which slid into the blade. This meant that we could adjust the length up to about 10 cm and thereby change the gearing. By shortening the oar, less power was required. On the other hand, it also meant we would sacrifice performance to some extent. However, we were now heading west with the wind and current, so that would help improve our speed. Having tried different adjustments, we settled on a 1.5 cm adjustment to our base position. That seemed to work well. It was slightly easier to row and there was no marked drop in performance.

Like the oars, the rudder did not take much adjustment either. Unless we needed to use the rudder to bring *Yantu* back on course when she slipped down a wave and turned sideways, we hardly made any adjustment to its position once we were rowing on course.

When we changed shifts, the person who was coming off shift would release the steering lines from the cleats next to his rowing position after the other person had fastened the steering line into the cleats next to his row station. Once the resting person was inside the cabin and had lain down, the person rowing would then do very slight adjustments to the rudder's position — maybe 1-2 mm — and then cleat the steering lines off again. Unless there was a change in wind or wave pattern, or the person inside the cabin moved, the whole shift could be rowed without further adjustment.

Although *Yantu* was seven metres long, where the resting person was in the cabin made a big difference to her stability. Sometimes we would have to yell to each other to move right or left in the cabin in order to trim the boat. In addition, the sleeping cabin was aft and when we climbed into it the end got heavier causing the front to rise out of the water.

We had been eating our way backwards in the boat, which meant we had removed about 60 kg of food from the front storage cabin. *Yantu's* front was therefore sticking up in the air, particularly when I was in the aft cabin sleeping. When Sun Haibin was sitting on his rowing station, he had to lift his arms high on his chest before the oars reached the water. This was not a comfortable position and something had to be done. After discussing it for a while, we decided to move the life-raft out of the well in the cockpit and put it into the front storage cabin. This had two immediate benefits. The boat was rebalanced and it was a lot more comfortable to stand in the well because there was now more space.

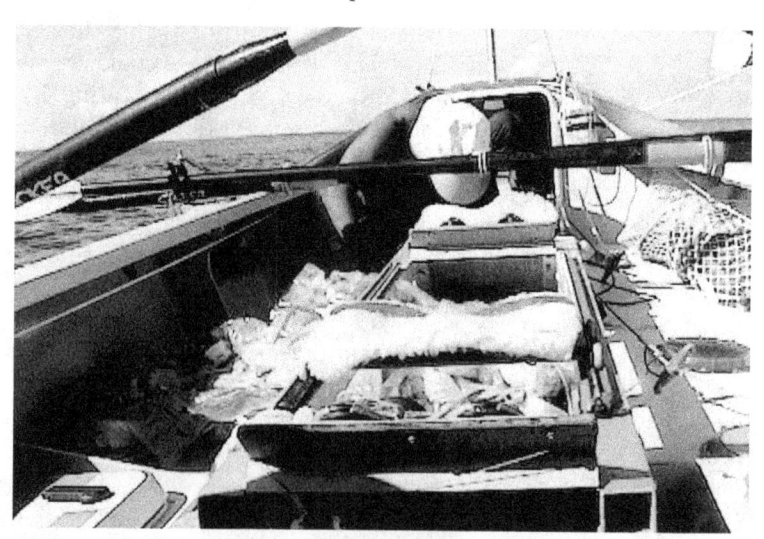

Rebalancing the boat by moving food around.

We never solved the problem of dampness in the cabin. As we worked our way south, it had become hotter and during daytime the temperature reached 37 degrees Celsius in the cabin. It was unbearably hot, but this could be relieved somewhat by opening the hatch on the cabin roof. The breeze would then pass in through the top hatch and out through the entrance hatch, which gave a cooling effect. However, the waves were coming from behind and when sleeping it was therefore important to keep our ears tuned for the sound of an approaching breaking wave. Once heard, the person in the cabin had just enough time to reach up and close the roof hatch. After the wave had hit, we would then open the hatch again. This was of course not fool proof and even though the person outside also kept a sharp lookout and yelled to alarm the person inside, sometimes we were too late and five or six litres of seawater would pour into the cabin. A lot of cursing could then be heard, water was bailed out, the cabin mopped as dry as possible and sleep resumed. Standard procedure!

The heat was also bad outside. Every time we got on watch, we would apply sun lotion and refill our water bottles before we started rowing. As it became hotter, our rowing outfit was reduced. During the day we wore a large brimmed hat, sunglasses, a long-sleeved blouse and that was it. No shorts. Rowing with a naked butt significantly reduced the amount of seat problems we developed. It was a trick we picked up from Rob Hamill's book. Rob had found it quite difficult to broach the concept to his rowing partner Phil out of fear of being taken for a queer. However, now that the idea was out of the bag, and was considered a critical success factor, it was easy for us to take on board. As a student Sun Haibin was living in a 20-square-metre dormitory with five other guys, so he was used to not having any privacy and from the changing rooms at the Yacht Club it was the same for me. But I guess it did not hurt we both knew the other had a girlfriend! We did discuss homosexuality and Sun Haibin, having been taught that it is a mental disease, was surprised when I told him that gay couples could legally marry in Denmark.

At night we rowed completely starkers, except that I wore glasses. Originally, I had not worn glasses at night, but after waking Sun Haibin up to get his confirmation that a ship was approaching from the horizon, only to be told it was the moon rising, I decided that glasses might not be a bad thing after all!

Rowing the 10–12 noon shift was the toughest as the sun was at its worst. We were sweating buckets, but we kept going as we were used to the heat from practising in Hong Kong. The shift was almost like rowing in a trance. The heat made it impossible to focus on anything and the mind simply faded out and tuned into the humming of the pistons from the water-maker, which we ran now ran from 7 to 12 every morning. The water-maker continued to work beautifully. Every two hours we would check the air trap. Depending on how rough the sea was, we would either have to clear out air every two hours or we could row for days before doing it.

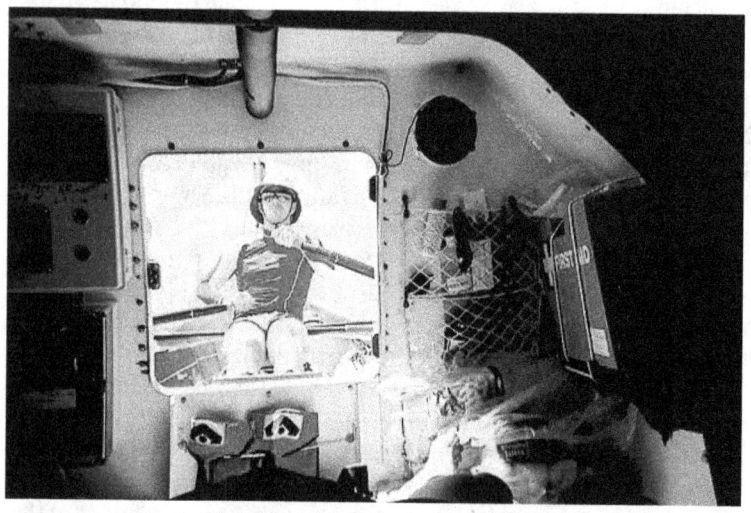

Another day in the office. Snapshot taken by Sun Haibin resting in the cabin out on me rowing.

Log entry — Day 22 — 29th October
Wind force 2-3 NNE. Waves 1-3 metres. Barometer 1013. Cloudy.
Great run last night — 54 nautical miles with little effort.
Thought about rebuilding my mom's summer house on Læsø more than 10 times while on watch. Never got it right and got tired of trying. Must find something different to think about.
The top hinge on the entry hatch into the sleeping cabin has broken. This happened before in Hong Kong. Must call Duthie at the Yacht Club to find out how they repaired it.
Sun Haibin repaired his seat this morning. He took some padding from the sleeping cabin and put it on the seat, but still not OK.
Dreamt Kate called me and complained loudly when Sun Haibin called me to go on watch. I though it was very rude of him to

interrupt my conversation. It took me a while to figure out I had been dreaming.
Called Véronique later in the day. We are now in 20th position.

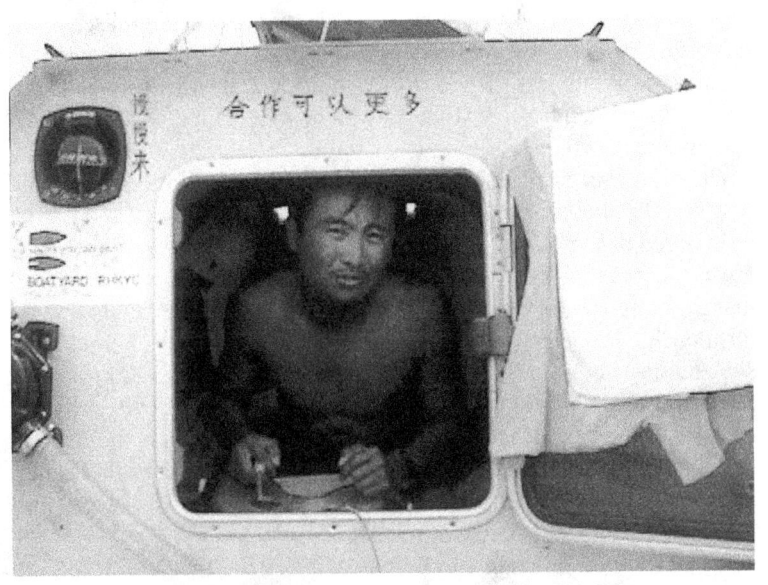

Sun Haibin using a file to reshape his rowing seat. The holes for the sit-bones were too far apart for his Asian behind.

It was a very interesting experience to observe what happened to my thought process as we rowed. I had made a promise to myself that I would only think positive thoughts when at sea. Because I hardly got any new inputs from the outside world, my mind would analyse the same topic for days on end. It might sound bizarre, but sometimes it was lonely being on board *Yantu*, despite sharing less than 10 square metres of space with Sun Haibin. We basically only saw each other for about 20 minutes at each watch change. Apart from that one was resting in the cabin while the other was outside rowing. Sometimes when the loneliness got too hard, we would ask the other person to stay up to talk. We swapped many stories as a result. One of my favourite anecdotes was about Sun Haibin growing up in Xinjiang. It came about as a result of an argument we had.

It is very common for Chinese people to fuss over each other to demonstrate that they care. Sun Haibin therefore frequently told me to put on sun lotion, drink more, rest more, put on a shirt to stay warm, make sure to brush my teeth; which all made me feel he was treating me like a child. One day I couldn't take it any more and lost my temper: 'For Christ's sake Sun Haibin. If I had wanted to row across the Atlantic with my mother I would have done so! I can take care of myself. Back off!'

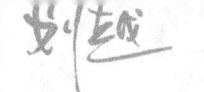

He did not know what had hit him. We then had a long discussion about fussing. To him it was very polite. To me it was very annoying. A classic case of cultural differences! We quickly made up again and Sun Haibin then told me a story that showed how difficult it was for him to stop fussing.

When he was growing up in Xinjiang his family used to have a pig, which ran around freely. Sun Haibin liked the pig very much and it was considered part of the family. It even came when it was called. One day, it was lying against a low wall outside the house. Sun Haibin decided to make it more comfortable and covered it with straw. Later on in the afternoon Sun Haibin's mother could not find the pig. To get a better view she climbed onto the low wall and called the pig. Nothing happened. She called again. Still nothing happened, so she jumped down onto the straw in order to walk back into the house. As she landed on the straw a lot happened! The pig yelled in pain and scrambled to its feet, knocking over Sun Haibin's mother. She immediately knew who was to blame and Sun Haibin was sent to bed without supper.

After I stopped laughing, Sun Haibin said: 'You see, that is just the kind of person I am.'

Sometimes I also annoyed Sun Haibin. Danish people do not like to sound too optimistic out of fear of attracting bad luck. We find it better to understate. Sometimes, if I said something very positive like: 'The wind and waves are great today. We will easily make 60 nautical miles,' alarm bells would go off in my head and I would immediately qualify the sentence with 'if we don't get hit by a ship'. Sun Haibin found that very hard to take. 'Why can't you just leave it sounding positive?' he would ask.

In fact, the only things we regularly argued about was his fussing over me, and my qualifying of positive statements. However hard we tried, neither of us could shake our habit. In total we argued maybe three or four times because of this, but never seriously and we made up within half an hour. We were lucky to share the same type of humour. Véronique observed in a supporter update from Tenerife: 'Christian and Sun Haibin get on well, they joke a lot, even Christian's Danish humour translated into Chinese seems to make Sun Haibin laugh....' Humour was a key success factor for us.

Heading west

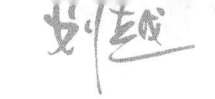

We only argued a few times and never seriously.

> *Log entry — Day 24 — 31st October*
> Wind force 2-3 NNE, increasing to 4-5. Waves 1-3 metres. Barometer 1012. Cloudy.
> Sun Haibin cleaned the bottom of the boat and we got about 0.2 knots from that. Rowed together for one hour. Total of 60 nautical miles in the past 24 hours! A little tired as we are trying to improve progress, but it feels good to cover so much ground. Talked to Duthie last night about the hatch. We will have to leave it. Duthie's advice: 'If it gets really rough both of you can always stay outside in the cockpit for a few days. You can then use duct tape to seal the hatch. In that way *Yantu* will still be self righting if you capsize.' Great advice....

Although the compromised hatch was a safety hazard, we decided it was not that serious. We had already been in very bad weather and had never been capsized, in spite of being flipped onto the side a few times. Because the waves generally broke over the back of *Yantu*, the aft cabin acted as a shield from the water and the cabin hatch would therefore not get a direct hit. Subsequently, it was unlikely we would get much water through that hatch, even in very rough weather.

Log entry — Day 25 — 1ˢᵗ November
Wind force 1-2. Waves 1-2 metres. Barometer 1012. Sunny.
Difficult to get speed into the boat due to wind and wave direction.
Had a talk about strategy and made a resolve to set daily distance targets. It can never be less than 52 nautical miles. We can get speed into the boat. Our bodies are getting used to the workload. Rowed together for about three hours to make the 52 nautical miles.
We threw three unused gas bottles overboard as well as 10 days' worth of food. Made the boat about 30 kg lighter, but we covered only 52 nautical miles.

Sun Haibin and I were both very competitive. We never arrived late on watch because we did not want one person to row more than the other. If the person on watch failed to wake the resting person when the two hours were up, and it had not been agreed to let the person sleep, the resting person got annoyed. We were there to row our 50 percent and the other person had no right to interfere with that! We both had personal pride as athletes, but I think we also had some sense of national pride. We were not going to let the other person ferry us across.

Once we rounded the waypoint and got into a rhythm, we both wanted to try and make our target of crossing in 56 days and finish in the top 10. Earlier on in the race when our performance started to slide we had reluctantly agreed to give up on the target and tried to be cheerful about it.

Although never openly discussed, we both knew in our hearts that when we had agreed not to care about our placing in the race, we had not been honest with each other. We would not be happy with just making it across. Success was defined as reaching the stated goals. We calculated that an average of 52 nautical miles would allow us to do that, so we set that as our daily minimum. We even went further. Every morning, when the 24 hours were up, we would set a target for the next day. Depending on how we felt it could be 52 nautical miles, but sometimes when the weather was good and we were feeling great, we would go for 60. Once the target had been agreed, we would pull out all the stops in order to achieve it. To allow us more time to row, we stopped eating meals together. The person cooking would eat first and then start rowing. The other person would then stop rowing, eat, wash the dishes and rest. However, sometimes, we did eat the night meal together, because it often consisted of instant soup noodles, which were too difficult for one person to pour into our eating containers.

Heading west

The consequence of setting targets was that we became obsessed with mileage. When we changed shifts we would ask the person about to come on duty how many nautical miles we had rowed in the past two hours. It had to be at least four, but five was better. Because the GPS only showed whole nautical miles, you could row 4.9 miles, but the display would only show 4. The other person would then come on and maybe only row 4.1 miles, but the display would show 5 miles. Every two hours it felt as if we were sitting an exam, but it kept us focused.

On 2nd November I also started calculating the daily average mileage required to reach Barbados by 3rd December. That day it was 50.6 nautical miles. It steadily decreased over the next weeks and by 16th November the required average was 49.3 nautical miles. It felt good to look at the declining trend of the numbers. Maybe we would even make it before 3rd December!

We also motivated ourselves by changing the frequency with which I plotted our position on the chart. I had plotted the position every day when we started on a new 24-hour period and then showed the chart to Sun Haibin. Due to the scale of the chart each day's progress only translated into about one cm on the 120 cm wide chart. Sun Haibin was always complaining he could not see any difference compared to the day before, which was discouraging. I therefore started plotting the distance only every four days. When I then showed the chart to Sun Haibin we had made it about five cm further west. This distance was easier to see and helped boost morale. We were getting across that chart!

It might seem like a crazy decision to throw out food. However, food equalled weight and the lighter the boat, the faster we would be able to row. When we decided to throw out food, we had already been at sea for close to one month and we had a good feel for what our daily distance was and therefore the amount of days left at sea. We still had food enough for a crossing of 65 days, nine days more than we expected to take, and that was without going on half rations, so the risk was limited.

Log entry — Day 26 — 2nd November
Wind force 1-2. Waves 1-2 metres. Barometer 1008. Sunny.
Last night at about 2AM it started raining and the wind turned against us. It has now stopped raining, but the wind is still against us. So much for the trade winds!
Extremely hot because of no wind coming through the top hatch. 37°C. Sweating buckets!
Managed 58 nautical miles, but it nearly killed us and I doubt we will make the planned 52 nautical miles tomorrow. After

Sun Haibin finishes the 19-21 shift we will decide on what tomorrow's distance target will be.

> *Log entry — Day 27 — 3rd November*
> Wind force 1-2 SSW. Waves 1-2 metres. Barometer 1011. Cloudy.
> Rowed only 40 nautical miles because of the wind and the waves coming from the side.
> Sun Haibin saw a ship during the night.
> Cleaned the bottom and I brushed my teeth and shaved. Sun Haibin had a shower.
> Now start on a new day....

The worst wave conditions we experienced were not big waves washing over us. It was small waves coming up from the south at very short intervals. This would happen from time to time for no apparent reason and when it did it was very frustrating. The boat would rock quickly from side to side making it very difficult to row. We would catch the oar blade in the waves on the return stroke and it would break our rhythm or we would bang the oar handle on our knees causing pain. Because of the rocking motion our butts suffered a great deal. It was like our butts were pieces of Parmesan cheese and the seat a cheese grater. Back and forth, side to side — our butts were being ground down. The wave set rarely lasted for more than 10 minutes and then it would disappear without notice just as quickly and unannounced as it had arrived. When these waves arrived the best thing to do was to stop rowing until they had passed, otherwise the pain and frustration could be excruciating.

> *Log entry — Day 28 — 4th November*
> Wind force 1-2 ENE. Waves 1 metre. Barometer 1011. Sunny.
> Wind finally turned behind us again and we managed to cover our agreed minimum of 52 nautical miles. Agreed 56 nautical miles for tomorrow's target. Hot day, but good progress. Called Véronique and we are now 14th!
> Had my weekly shower and it was great! Cut the nails on my fingers right down so they don't scratch my knees when I row. My butt is not too sore. If our butts go, then we are finished.
> The Kiwis have only 700 nautical miles left to go and it looks like they will break the 41 day record. Good on them!

Heading west

> *Log entry — Day 29 — 5th November*
> Wind force 2-3 NNE. Waves 1-3 metres. Barometer 1010. Sunny.
> Wind back to normal and waves are great. We covered 59 nautical miles — not bad! We are now in 12th position.
> Threw out the rest of the food for week 10 & 11 — about 10 kg.
> Washed our butts and also the lambskins we sit on. Hopefully our behinds will get better soon.
> Sun Haibin's water bottle broke today. It had been thrown too many times across the cockpit. Repaired it with Blu Tack, a plastic bag and duct tape.
> Rowed the 16-18 shift and for the first time felt that *Yantu* was light — I hope the feeling stays!
> Sun Haibin called Zhang Jian and Cao Xinxin.

One day Sun Haibin stayed up to talk to me about his army training. His parents were originally from Henan Province, but had been sent to Xinjiang during the Cultural Revolution. When Sun Haibin was a middle school student they had been allowed to return to Henan. There was not much going on in Henan for a guy like Sun Haibin so he decided to join the People's Liberation Army as a professional marathon runner. In China, in addition to the special sports institutes, the PLA is also used to nurture athletes. His idea was that once he joined the Henan PLA, he could apply for a transfer to Beijing. It almost worked out. He was transferred to what he was told was Beijing, but in reality it was to Shijiazhuang, a city 250 km south of Beijing and of even less interest than Henan. He was now stuck. However, he knew that China's most prestigious army sports team, Unit 81 in Beijing, went talent scouting across China from time to time. He decided his ticket out of Shijiazhuang was to get into Unit 81 and he started training for the 1500 metres, as he knew this was the distance the talent scouts would be interested in when they arrived later that year. Without permission he started getting up at five o'clock in the morning to run for an hour before the rest of his unit got up. It would be dark outside and the only place to run was a 200-metre circular course. Because of its limited size he would often slip and bang into the surrounding barrier, prompting the guard on watch to yell 'Who goes there?' To which Sun Haibin would reply: 'It is just me running.' Rumours soon started to circulate that there was a mentally disturbed soldier in one of the army units....

Sometimes they would have to stand to attention for hours on end with all muscles tensed. Their commanding officer would then walk behind them and randomly tap the back of someone's knee. If the knee shot forward it meant the person was not tensing his muscles and he would be punished as a result. Even if they farted they would be punished. The officer would yell: 'Who told you to fart? If you are hiding from the enemy and you fart they will find you. Only I decide when you can fart!' Listening to these stories I had to ask Sun Haibin which was harder, his army training or rowing the across Atlantic.

Sun Haibin in military uniform trying hard not to fart.

'Without a doubt it is rowing across the Atlantic. The two don't even compare,' the reply came back.

His army unit did not put him forward for the 1500 metres race when the talent scouts from Beijing arrived, but Sun Haibin decided to take his chances and went directly to the referee, and told him he had to run the race because he wanted to get to Beijing. The referee allowed Sun Haibin to run. He left the other competitors in the dust and was accepted into Unit 81.

Once in Beijing he was told to train as a triathlete. He did this for some years, but became increasingly unhappy with the coaching. Eventually he told his coach he reckoned he would perform better if he coached himself. This caused a big stir, but after signing a document, including thumbprints, to the effect that if he died during training, it was no one's fault but his own, he was allowed to train by himself. He then went on to beat the athletes the coach was training, but understandably his flamboyant move had a career limiting effect. He therefore applied to Beijing Sports University and once accepted he left the army. When I met Sun Haibin, he had been studying at the university for three years and he was in his final year. He had now again gone against the mainstream. His final exams would have to wait. Sun Haibin certainly was a determined and focused guy, but on the 6th of November he was not his old self.

November 6th was Sun Haibin's 26th birthday. The weather was beautifully sunny with the usual NNE wind blowing a force 1-2 and the waves running at a friendly 1-2 metres high. I sang *Happy Birthday* for him and promised him we would see a dolphin that day in honour of his birthday. I also presented him with a gift of two litres of fresh water so he could have a birthday shower, but not even that made him happy. After breakfast, he rowed his morning shift, but kept complaining about the speed. However hard he tried he could not get it above 1.6 knots. I had been doing 2.5 to 3.0 knots. By the time his shift was over, he was in a foul mood.

Then something great happened. Suddenly a large pod of pilot whales appeared around us. For two hours, whale after whale passed us on exactly the same course as ours. It was as if they were also racing to Barbados. *Yantu* was in the middle of a host of large whale fins and we took out the video camera to shoot some footage. The sight was breathtaking.

'I told you we were going to see big fish today,' I said. 'I really had to use a lot of *guanxi* (Chinese for "connections") to pull off this birthday display.'

He was still not smiling. I suddenly understood that he was homesick.

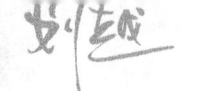

'Sun Haibin, go into the cabin and call all your friends, and I mean all of them. Cao Xinxin, Zhang Jian, and anyone else you normally celebrate your birthday with,' I said.

'Don't be stupid. I don't need to do that. It is expensive,' Sun Haibin replied.

'Do it,' I insisted.

He disappeared into the cabin to make his calls. It was a very different Sun Haibin who reappeared an hour later. He was now beaming from ear to ear.

'Ahh, Christian. That was the greatest thing. I even called some of my old army buddies, who told me they wished they could be here to celebrate with me,' he explained, and suddenly tears welled up in his eyes. It took him a minute to compose himself, then he said: 'I now understand the emotion you felt on the day the race started.'

Sun Haibin calling his friends and family on his birthday.

We talked some more and then we changed over. Sun Haibin started rowing. I looked at the GPS and said: 'You are easily doing 2.5 to 3.0 knots now. There was nothing wrong with the boat before. It was just you feeling sad.'

Sun Haibin was steaming ahead and we kept talking. Through Véronique I had arranged for Sun Haibin's new buddy, Tom from *American Star*, to call him on his birthday, but we were so engaged in

Heading west

our conversation that we forgot to switch the phone on. That was a great shame, but nothing could now ruin the day. We had fried rice that evening, which was Sun Haibin's favourite, and when it got dark I gave him a white flare to set off. I knew he really wanted to try that. To finish off the birthday, I took out my birthday present for him. It was a key ring with a blue dolphin in it. Véronique had bought it in Tenerife. I had given her the instructions to get something 'light, small and compact' so the weight would not slow down the boat and she had chosen an excellent present against these criteria. As Sun Haibin opened the present, I said: 'There you go. I told you we would see a dolphin today.'

When I next came into the cabin to rest I saw that Sun Haibin had hung the dolphin from the cabin roof hatch. It was pointing towards Barbados and rocking gently with the boat. It stayed there for the rest of the trip, encouraging us on. Every time we looked at it, it made us happy.

We were on a high for the next few days, but then the weather turned on us.

Log entry — Day 33 — 9th November
Wind force 1-3 SW to NNE. Waves 1 metre. Barometer 1010. Sunny, 37°C.
Finally got a lot of wind and waves from the right direction, but not for long. There was a lot of swell, which made the scenery more interesting because we could see far from the top.
Sun Haibin repaired his seat. I cleaned the cockpit well. We moved food to rebalance the boat. Now the four side compartments next to the rowing stations are empty.
Did 56 nautical miles. I constantly row a little over 2 knots. Tough though.
My big toe is getting a little worse, but not a major issue.
For the first time the battery dropped below 12.4 Volts. Hopefully no problem. It is now back at 13 Volts.
We are apparently still in 13th position.

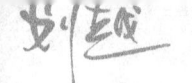

Log entry — Day 34 — 10th November
Wind force 1-3 SW. Waves 1 metre, messy. Barometer 1012. Sunny, 37°C.
Rested together for three hours this morning. Both completely tired. Wind against us — did only 40 nautical miles. Rain from the south, but not a lot. I tell you there is still a long way to go....
Lots of rain in the afternoon. Made me feel cool and much better.

Log entry — Day 35 — 11th November
Wind force 1-3 SW. Waves 1-2 metres. Barometer 1012. Rain, 35°C.
Lots of rain last night. Funny thing happened — I had one complete watch without rain, the only two hours of the night without rain!
Cleaned the bottom this morning. We both swam and showered. Making steady progress. Did 53 nautical miles.

Log entry — Day 36 — 12th November
Wind force 1-2 SSW. Waves 1-2 metres. Barometer 1012. Sunny, 37°C.
Sun Haibin surprised me this morning by saying he wanted to stop setting daily targets. We eventually agreed to try and beat 12th position. He is now OK again. It turned out it is the rain, not the daily targets, which is getting to him.
Very hot to row. No wind. Did 52 nautical miles. Target for tomorrow is 56 nautical miles.
We passed Tom and John in *American Star*. This is a major win for us. We joked in Tenerife about who would get there first and it now looks like it is going to be us. It feels good!

Log entry — Day 37 — 13th November
Wind force 2-3 SSW, later 1-2 NE. Waves 1-2 metres. Barometer 1009. Sunny, 27°C.
Rowed 56 nautical miles! Now in 11th position. We also passed through the 1,000 nautical miles to go barrier. Great mood on board!

Sun Haibin and I were increasingly living in our own universe and focusing on two things only — speed and position. Véronique could feel this when I called her. In her diary to me she wrote:

Sometime I do think this is a really selfish thing you are doing, even though I want nothing else than for you to do it. I would like you to understand how hard it is for me to support you while hiding my sadness, my fears, my worries. It sometimes feels like such a one-way thing: I am here for you, but you cannot be there for me. Most of the time it is OK, but today I experience it as hard on me, unfair, and I wish this was not going to take another month. To know that you are having a hard, painful, and uncomfortable time makes it even worse.

A few days later she continued:

It was good to speak to Sarah today [the girlfriend of Tom, rowing American Star] who is going through very similar emotions; too much work for the race, too many emotions not communicated over the phone, too little acknowledgement in return. We both agreed that even though we love you like crazy and are extremely proud of you, you are also all a bunch of selfish bastards!

Véronique fortunately also called my sailor friend Olav and talked to him. The next time I called him, he told me how Véronique was feeling and through him I arranged to send her flowers. Her diary reads:

Tonight when I came home there were flowers waiting for me. I could not believe it: roses from you with a lovely, loving note! This made me cry. You are so wonderful. I want you back soon and then I never want to be separated for that long again!

I was back in her good books, but I still had to row 1,000 nautical miles before we would be together again.

The last thousand miles

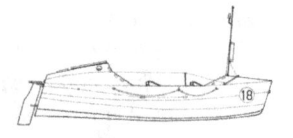

The wind that started on the 13th of October made the day shifts less punishing to row, but the 13th lived up to its ominous reputation that night.

After having eaten Mexican chilli con carne for dinner I went into the cabin to sleep. The wind was rushing down through the cabin roof hatch and was nice and cool. So cool that I caught a stomach cold and had to rush out and interrupt Sun Haibin's rowing to get onto the red toilet bucket. On the whole trip, apart from this one incident, neither of us suffered from a stomach upset, but that night I remember. When I had eventually finished, it was my turn to row and I settled into my rowing seat. Before going inside the cabin to rest, Sun Haibin told me he thought he had seen a ship far behind us. Sometimes, when he was on the top of a wave, he could see a light.

I started rowing, but could not see any light. I scanned the horizon behind us and eventually I spotted it, but it was very far away and nothing to worry about. Over the next two hours it came closer. It was now completely dark so the light was easy to see. We changed watch again.

I had not been in the cabin long before Sun Haibin asked me to come out to have a look. The ship was quite close now and we could see its silhouette against the sky.

'I think it is coming straight at us,' Sun Haibin said. I had a look and had to agree.

'I'll try and raise it on VHF channel 16, just to make sure it has seen us,' I replied and went into the cabin to get the radio. When I came back out, the ship was a lot closer and definitely coming straight at us.

The last thousand miles

'This is *Yantu* calling approaching vessel, over,' I called. No reply, just the hissing from the radio. I switched on the GPS and got the long and lat.

'This is *Yantu* calling vessel in approximate position 15 degrees and one minute North, 42 degrees West, over,' I called again. I released the call button and waited. Still nothing. Just hissing. The ship was now so close it was not funny any more. I called another three times in quick succession. Suddenly the VHF came to life. It was playing loud music.

'Hello?' a voice spoke over the music.

'You are about to ram us. Please turn either port or starboard.'

'But I can't see you on my radar.'

'Then look out your window.' A slight pause, then the voice came back.

'What the hell kind of a vessel are you guys?'

'We are a rowing boat. Now TURN!'

The ship turned. We relaxed. The voice came back on the radio.

'What are you guys doing? Are you OK?'

'Yes, we are part of a rowing race across the Atlantic, so please keep a look out. There are 36 of us out here!'

'Oh, I heard about you guys in Tenerife, but I never thought I would see you. Where are you heading?'

'We are going to Barbados. Should be there in about 20 days.'

'So am I, but I think I will be there in six.'

When the ship had turned it also slowed right down. It was now passing us about 100 metres away. It was a big sailing yacht going under motor, six or seven times the size of *Yantu*. Now that it was not about to hit us any more, it looked nice and modern. We talked for a while longer and then the captain put the throttle down and sped away. Soon we could only see its white top light. Sun Haibin and I looked at each other.

'Did you see that ship. I bet it had air-conditioning,' Sun Haibin started.

'A fridge and a proper bed, too,' I continued.

'Great showers and good food,' Sun Haibin fantasised.

203

The ship in the dark came too close for comfort.

'Only six days to go for him,' I said with a longing voice.

Up until that point we had been living in our own world. We had decided whether and when to contact the outside world through our satellite phone. Now we had been imposed upon by the yacht and it had rubbed all the luxuries we were deprived of in our face. It had also damaged our feeling of safety. Our mood hit rock bottom. For the rest of the night we were unable to get back into our rhythm. We had been destabilised.

Because of the change of direction of the wind from SSW to NE, the waves then started playing up and although they were not very big, they were messy and difficult to row in. To top it all off, when I rowed the morning shift a set of short waves from the south arrived rocking the boat violently from side to side. The sensible thing would have been to stop rowing, but we had had enough distractions. We needed to get back into that rhythm.

I am going to row through this, I promised myself, gritted my teeth and continued rowing.

The short waves kept coming and I kept going. They were not going to get the better of me. The set continued. I was hitting my hands on my knees and scratching my legs with my fingernails, but I kept going. The waves caused my butt to ache as if it was sliding from side to side over a washboard and, as my body rocked from side to side, the bottom of my rib cage started rubbing against the top of my pelvis with a grinding sound. The pain was excruciating and it had become personal. It was me against those waves and I was not going to give in. By the time Sun Haibin came out of the cabin I was in a terrible state.

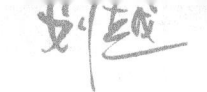

He took one look at the pain on my face. 'What happened?'

'Those fucking short waves from the south just kept coming for the whole shift. They couldn't just make it the normal 10-minute set. No, no. It had to be the whole shift. I was stubborn. I kept rowing. I am completely done. I need to rest.'

I went into the cabin and passed out. The pain was intense and I felt intensely stupid. I was unable to row my next shift so Sun Haibin rowed it for me and then he continued with his own shift after that. After five hours I was still in pain, but fit enough to go outside to make lunch.

I poured the water into the pressure cooker and then I reached for the lighter to light the stove. It was out of gas, so I went into the cabin to get the spare one. Back outside to the stove. I flipped the lighter. Nothing happened. I held it up against the light. Also out of gas! I then opened the compartment under my rowing seat and took out the aerosol lighter re-charger we had bought in Tenerife in case this happened. Flipped the switch a second time. Still nothing happened. This was unexpected. Again I held the lighter up to the light. There was no gas in it. I shook the aerosol can. It was empty too! The gas had leaked and evaporated because of the high temperature. We were mid-Atlantic with 20 days to go and no ability to cook. Major bad news.

'Sun Haibin, we are out of lighter gas. It has all evaporated, even the spare aerosol can. It looks like we will be eating cold freeze-dried food for the rest of the trip.'

Sun Haibin stopped rowing in disbelief and came to join me in the cockpit. He examined the lighters and the empty aerosol can.

'I think I might be able to get the spark mechanism out of the lighter. When I was a kid in Xinjiang we used to take it out to play with. It was great for giving someone an electric shock.'

He then went inside the cabin to take the lighter apart and I started rowing. Some time later he reappeared with a 1.5 cm long brick-shaped iron thing between his thumb and index finger. He came up to my rowing station.

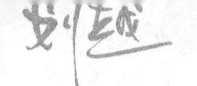

'Hold out your hand.' I held out my hand. He then put the metal thing next to my skin and pressed the ends together. There was a click followed by a sharp pain. I instantly retracted my hand.

'It works! Now let's see if it can light the stove,' Sun Haibin laughed and turned to switch on the gas.

Many clicks later the stove finally lit and a large flame licked up around Sun Haibin's hand. He moved it quickly and started shaking it, but there was still a smile on his face. The stove was lit and we could cook. For the rest of the trip we would make bets about how many clicks were needed to light the stove. If we managed it in five it was considered lucky because the size of the flame that flared up around your hand was then still quite small. Needless to say we took very good care of the little sparking mechanism and it worked fine until we reached Barbados.

After lunch, Sun Haibin went into the cabin to rest.

'Christian, we have another problem,' his voice sounded from inside the cabin.

'What is it?'

'The hinges on the cabin roof hatch are broken.'

'What! Are you sure?'

'Come and see for yourself.' I stopped rowing and joined him in the cabin. The hinges were indeed broken. Major, major bad news!

The cabin roof hatch was the most exposed to the elements because waves came washing over from the back of the boat. If a wave came over us now it would rush right into the cabin and fill it. This was a very serious safety hazard. The weather right now was fine, but we still had three weeks to go and we could not count on it being fine all the way, particularly since a hurricane, Olga, was reported to be forming in the Caribbean. We now had two broken hatches into our sleeping cabin. It was not safe to continue.

Can it get any worse? I thought. The experience hit us deeply.

Sun Haibin and I discussed what to do and eventually we decided to use some screws to try and make replacement hinges. We took out the old hinge mechanism. It was made of plastic with a small steel pin inside it and looked more like it belonged in a Lego toy kit rather than onboard a boat. The replacement hinges kind of worked. If the hatch now got hit by a wave it would stay attached, but it would leak a lot. Some of the leakage could be controlled by using duct tape. Was this safe enough to continue? We debated it for a while. Basically, unless we capsized or got hit by a monster wave we would be OK.

The last thousand miles

So far we had experienced neither. Even if we were to capsize and had to abandon ship, we still had our life-raft and EPIRB. We agreed to take a calculated risk and row on. The idea of having to give up because of three lousy hinges was unbearable.

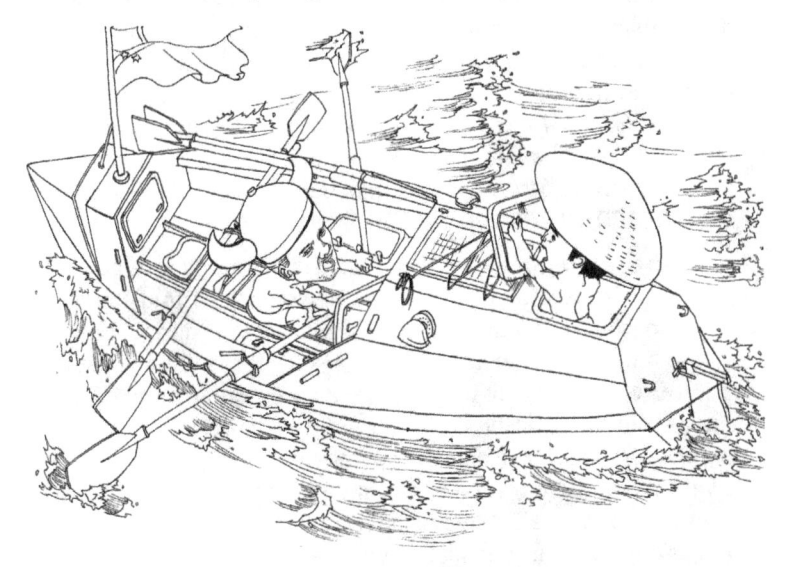

The broken hatch hinges were a major safety hazard.

I called Duthie at the Royal Hong Kong Yacht Club to get his opinion. After initially being somewhat dismissive when we called about the solar panels, Duthie had become considerably friendlier since we started regaining places. He was now convinced that we were there to go the distance. He promised to get in touch with Lewmar, the hatch manufacturer, to see if they had any advice. I called him back later.

'You know what they told me?' Duthie reported with a tone of disbelief.

'No, what did they say?'

'They told me the best thing for you to do was to retire from the race and get spares from the safety vessel,' he replied.

'Well, what did you reply?'

'Not fucking likely!'

Duthie was right behind us!

A few days later Véronique received a letter from the Challenge Business, sent to me in Hong Kong and subsequently forwarded to her in Munich by my Hong Kong flatmate. The envelope was dated

22nd August 2001 and the leaflet enclosed was dated 20th July 2001. It read:

Product Safety Notice for Lewmar Low & Medium Profile Hatches

Problem: It is possible for one or more of the hinge pins to break, usually as a result of incorrect adjustment. This may not be apparent as the hinge pins have a small stainless steel pin within them to retain the hatch lid should the hinge pin fail. It is *essential for the safety and security of the yacht* that the hinge pins of all hatches shown above are checked for damage in line with the attached sheet labelled **HINGE CHECK FOR LOW & MEDIUM PROFILE HATCH.**

Needless to say Véronique did not particularly enjoy receiving this letter. Duthie subsequently contacted Lewmar, who told him they had sent spare parts to the Challenge Business in order for them to check the hinges in Tenerife prior to the start. If the Challenge Business and Lewmar knew of the problem before the race started, then why had it not been picked up during scrutineering?

I believe the answer lies in the postage on the letter. Most of the competitors were from the UK and my letter must have gone into the domestic pile when stamped, so it was under-franked and went to Hong Kong by surface mail. I can only assume that the Challenge Business thought I had received the letter prior to arriving in Tenerife and therefore believed any hatch issue had already been taken care of by the time they did the scrutineering. Sometimes shit just happens and at least Chay saved a few pennies on stamps!

Log entry — Day 39 — 15th November
Wind force 1-2 NE. Waves 1-2 metres. Barometer 1010. Cloudy, 37°C.
Good speed in the boat. Hopefully back to normal. Did 47 nautical miles.

Log entry — Day 40 — 16th November
Wind force 2-3 NE, later SW 1-2. Waves 1-2 metres. Barometer 1009. Sunny, 35°C.
Great wind and waves. Did 57 nautical miles with little effort. Wind shifted in the afternoon, but still good speed. My butt is OK today — touch wood.
Getting back into the swing of things....

The last thousand miles

I called Véronique who read me an e-mail she had received from the Chinese students at Atlantic College:

Dear Christian and Sun Haibin,
How are you doing? We hope everything goes well in your rowing across the Atlantic. We feel very grateful to you for doing that!
We are the three students from mainland of China in Atlantic College. One of us is called Li Zhe, is in the second year, comes from Guangzhou; the other two first-years called Jin Zhou and Maha Tiri, comes from Guangzhou and Sichuan respectively.
We are having an absolutely excellent experience in Atlantic College. We just can't imagine, before our coming, there is such an amazing school. We are enjoying ourselves in the IB [International Baccalaureate] course, the service programme (two of us are lifeguard while the other is coastguard), the activity, and perfect opportunity to live with, to know, to understand friends from all over the world. It is the thing we have few chances to do in our motherland. On the other hand, as students from an old country which is rich in its oriental traditions, cultures, and making enormous progress in recent years, we are proud of ourselves and eager to let people from the rest of the world know about us, about our treasures of history, philosophy, wisdom, culture, and about the great change which is taking place nowadays.
We try our best to learn nice things from others, and to present what we know about our country through daily life, activity, national evening and various other opportunities. We feel really grateful to the people who provide us scholarship, by which we can have such a great, completely different experience during our lives.
However, as the main part of the country, which raise one fifth of the world's population, there are only three representatives in Atlantic College. How we wish we could have more youths from our motherland here, share the experience with the students from over 70 countries. It is also the reason we are absolutely moved and grateful to you when we heard that you are rowing across the Atlantic to raise money for another Chinese student to come to this marvellous college. Thanks a lot!
We wish the best of luck during the great trip and the best future!
Yours sincerely,
Li Zhi, Jin Zhou, Maha Tiri

That cheered us up and gave us renewed energy!

> *Log entry — Day 41 — 17th November*
> Wind force 1-3 NE, later none. Waves 1-3 metres. Barometer 1009. Sunny, 37°C.
> Trades died as predicted by the weather forecast. Sun Haibin got burned by plankton when cleaning the bottom of the boat. He was in bad spirits. I talked and joked with him a lot and that cheered him up. We are good at encouraging each other!
> Rowed 54 miles.

Sun Haibin was OK on top of the ocean, but he was not always comfortable swimming in it. He was used to swimming pools and always spent a lot of time looking for sharks before he got in. He was therefore pretty badly shaken when the almost invisible plankton stung him. He did not know that such things existed and flew straight back up into the boat. Once I had told him what it was and that it was not dangerous, he was fine with it. We rowed on for a while to get away from the plankton and then he jumped back in and finished the cleaning. This was the only time he got stung.

The real danger with cleaning the bottom of the boat was the waves. When the waves were more than one to two metres high it was too dangerous. If *Yantu* pitched and then came down on top of him, he would have been seriously hurt. Sometimes we had to delay cleaning the bottom for one or two days because of this.

> *Log entry — Day 42 — 18th November*
> Wind force 1-3 SE. Waves 1 metre. Barometer 1010. Sunny, 35°C.
> Rowed 52 nautical miles. Slow pace.
> Took food from the front of the boat.
> The Kiwis finished today in 42 days. They missed Rob Hamill and Phil Stubb's 1997 record by one day! They must feel bombed out!

> *Log entry — Day 43 — 19th November*
> Wind force 0-2 NE. Waves 1 metre. Barometer 1009. Sunny, 37°C.
> Rowed slowly during the night — about 1.5 knots — due to lack of wind and waves.
> Skipped the 7-9AM watch and we both slept. Only covered 42 nautical miles.

The last thousand miles

> Wind picked up a bit in the afternoon and we made better progress.
> Changed the time on the boat from GMT to Barbados time. Makes more sense with the hours for meals. Also, it is good psychologically — we are getting there!
> Shaved in the afternoon and then showered. Also repaired squeaking rowing seats.

Because we were travelling west with the sun, every day was slightly longer than if we had stayed stationary. Over time, our watches set on Greenwich Mean Time therefore stopped correlating with the day we were experiencing. Our watches would show 8AM and so time for breakfast, but when we looked at the sun, it was at the zenith: midday. For a while we were having lunch at supper and supper for the midnight snack and the midnight snack for breakfast, which became increasingly annoying.

After we changed our clocks to Barbados time, the world was more normal. Sunrise was now slightly before 8AM and at lunchtime the sun was overhead.

> *Log entry — Day 44 — 20th November*
> Wind force 0-2 NE. Waves 1 metre. Barometer 1009. Sunny, 27°C.
> Did 52 nautical miles last night.
> Managed to get speed up to 2.7 knots, but the direction was too far north. Once back on course speed was 1.8 to 2.0 knots.
> Very painful boil on my left butt cheek.
> A ship passed close by last night — did not try to contact it.

After our near-collision experience and the intrusion of our privacy, we agreed not to contact other ships unless it was with a clear purpose. We did not want to ruin our rhythm again. In total we saw six or seven ships during the crossing, but only talked to three of them, one of which was the Challenge Business rescue vessel.

> *Log entry — Day 45 — 21st November*
> Wind force 2-3 NE. Waves 1-3 metres. Barometer 1008. Sunny w. clouds, 27°C.
> Did 51 nautical miles. Trade wind is back, but not yet regular.
> My butt is better — managed to row five shifts without pain. *Great!*
> Sun Haibin has had a nosebleed for the past two shifts — we are really pushing it!

Our required average daily distance to get to Barbados for 3rd December was now 48.6 nautical miles. We were even starting to think that we might get there one or two days early. But the good progress was coming at a price.

When we came on watch and started rowing the joints in our hands were cold. It took about five minutes for them to warm up. During those initial five minutes, it felt like the oar handle was covered in broken glass and each pull on the oar was painful. Our hands were also full of blisters and hard skin. I did not mind this, but Sun Haibin had problems with his skin cracking and then when salt water or sweat got in it would be very painful. To compensate for my lack of cracked skin, I had an ingrown toenail, which required cleaning every day. My right rowing shoe was dark from the pus and blood. But none of the above hurt when compared to our butts.

Sun Haibin's showing off his blistered hands. Still nothing compared to the state of our butts.

Bob Wilson in Hong Kong had been absolutely right. Our sit-bones did indeed go through the holes in the seat, well assisted by the constant backward and forward movement from the rowing and the sideways motion from the rolling. Initially the problem was sweat and could be relieved by rowing butt naked.

The last thousand miles

But as time passed the sit-bones started getting bruised and they never got sufficient rest to recover properly in between shifts. They could therefore only get worse.

We tried everything to make the seats more comfortable. They were standard sculling seats covered with a layer of foam camping mat. But we had been sitting on the seats for so long that the foam had collapsed. We then tried cutting out foam from the cabin and placing it on top of the seats. That worked for about five days before the additional layer of foam also collapsed. We put a lambskin on top of the foam and got another five days of grace. We added another lambskin — another five days before the pain came back. We were out of lambskins and resorted to a folded towel. The towel was put between the two lambskins and we gained another five days before the pain came back. By the time we had finished constructing this princess-on-the-pea padding we were out of options, but still had about three weeks of rowing left to go. That was when we started taking painkillers, but it was still very difficult to ignore the pain. Particularly when we started getting boils. I vividly remember my first boil.

I had come off watch and went into the cabin to sit down. A sharp pain came up through my butt and I felt like I had sat down on a hard stone, but when I stood up to check there was nothing under me. The sharp stone was under my skin and every time I sat on it, it was extremely painful. I tried to look at it with a mirror, but it is not exactly an easy place to look at, particularly not when you are lying on your back in a small rowing boat rolling 45 degrees from side to side. Eventually I swallowed my pride, went outside, turned my butt towards Sun Haibin, who was rowing, and said:

'What is wrong with my ass?'

'Oh, you have a big pimple on it. It is very red and swollen,' the reply came back.

This became the beginning of an ungraceful ritual of calling out the state of each other's ass to each other.

'Today you have two new ones on the left cheek, and it looks like the big one on the right has burst....'

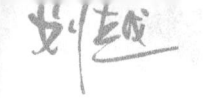

Getting a second opinion on the state of my behind.

A boil would become increasingly hard and painful and then after four or five days it would burst like a ripe zit and leave behind a crater once the pus was cleared away. They were very, very painful!

Preparing the rowing station became something close to a religious ritual. We would arrange the collapsed foam, then ever so carefully put down the first lambskin on top. Then the towel was folded with extreme care and added. Lastly, the second lambskin was placed to top it all off and gently smoothed to ensure no wrinkles were present. The seat was now ready.

Next came the challenge of actually sitting on it. Gingerly we would position ourselves over the seat and then sit down. One of two things would happen. Either the next two hours would be relatively painless, or it would hurt like crazy and the shift would be hell. It did not work to stand up and try again — that only made the pain worse.

Despite the pain, the sitting down became the highlight of each watch shift. What would happen? Bearable pain or hell? When it was hell the person would inevitably curse, which would cause the other person coming off watch to laugh, well knowing he would be the literal butt of the same joke two hours later.

The last thousand miles

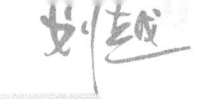

> *Log entry — Day 46 — 22nd November*
> Wind force 1-3 NE. Waves 2-4 metres. Barometer 1008. Sunny w. clouds, 27°C.
> We are now in 10th position! Good speed. Did 63 nautical miles last night.
> Cleaned the bottom of the boat.
> Called Duthie to get his view on whether to go to the waypoint south of Barbados. His comment: 'You guys just steer directly on the lights in the bar on Barbados.' He sounded really excited.

> *Log entry — Day 47 — 23rd November*
> Wind force 0-2 N. Waves 1-2 metres. Barometer 1007. Sunny, 27°C.
> Cool wind made rowing at noon OK.
> Covered 53 nautical miles — not bad!
> My butt hardly hurts now, which means I can pull 0.5 to 1.0 knots more — a big difference!
> Véronique said the hotel rooms in Barbados are confirmed. Can't wait.
> Had a celebration lunch of tinned mackerel in tomato sauce.
> I feel excited....

The pain never went away and it increasingly consumed us. Humour was called for.

'Sun Haibin, if I ever meet an ocean rower who tells me his butt did not hurt, I am going to beat him up badly.'

'Yah, maybe we can tie him behind a car and drag him on his butt over a dirt road with lots of sharp stones in it.'

'Yes, that would be good — and then we rub some salt in the wound.'

The punishments became more and more sadistic, but we never came up with a torture we felt would outdo our own pain!

Normally, our body clocks were very accurate and the first time we would look at our watch to see how much longer we had to row was after one hour and 45 minutes. However, when we sat down and had to row a shift in hell, the body clock was off. We knew this and would therefore try to wait longer than normal to look at the watch, but only to be disappointed. Only five minutes would have passed. We would let another long period of what felt like at least an hour pass and then look again. Only another 10 minutes! Shit! The hell shifts seemed to last forever.

That said, I was amazed at the speed with which our bodies were able to recover. After rowing a hell shift we would feel completely drained and unable to move from pain, and then after only two hours of rest we could do it all again.

> *Log entry — Day 48 — 24th November*
> Wind force 2-3 SW. Waves 1-2 metres. Barometer 1007. Sunny w. clouds, 27°C.
> Rowed into the wind, but easily did 2.5 knots because of the current. Later only 0.9 knots. Stopped rowing at 6AM — it was not worth the effort. Only covered 42 nautical miles.

Despite the pain, the last two weeks of the race were probably the best. We found a good rhythm and we slowly clawed back positions. We had had very little outside interference and our minds gradually became completely empty, as if we were meditating. The feeling of rowing and being so close to nature and surrounded by so much beauty was incredibly soothing. Although there was only the ocean and the sky to look at, it was never boring. At night we could see the smoke trails behind the shooting stars. Sometimes I spotted a white cap on a wave even when there was hardly any wind. I then knew a whale had come up to breathe. Another day we rowed past a leatherback turtle, which was sleeping on the surface. The clouds were moving in the sky and forming all kinds of shapes and sizes. I could spend a whole watch spotting figures in the clouds. One day I saw Mickey Mouse clear as day and I woke up Sun Haibin to share it with him, but by the time he got outside Mickey had changed into a cactus! As we got closer to Barbados, the sun at dawn and dusk was dramatic when it broke the cloud cover. The depression forming in the Caribbean was giving us plenty of clouds to look at and they moved fast across the sky in all directions.

> *Log entry — Day 49 — 25th November*
> Wind force 2-4 SW. Waves 1-3 metres. Barometer 1007. Rain then cloudy, 28°C.
> Stopped rowing. No point with the wind on the nose. Put out drift anchor. Went for a swim and cleaned the boat. Went fishing. Slept together in cabin. Good rest, if not comfortable. Only covered 5 nautical miles.

The last thousand miles

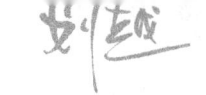

Until this day we had needed only 46 nautical miles per day to make it for Véronique's birthday and since we were rowing an average of 52 it looked like we would arrive early. However, after this day off the required average was back up to 52 nautical miles.

It was strange suddenly not to have to row. When that had happened in the past it was because we were inside the cabin being beaten up by breaking waves. Now it was a beautiful day, the sea was flat, and the sun was shining. We tied the fishing line to the boat and went swimming instead. We had been too busy rowing until then to even think of fishing. Because of the drift anchor we were not moving and the fishing line hung lazily down under the boat, so it was really a wasted effort. However, it gave us something to do.

There was no risk of *Yantu* drifting away given the weather conditions and the deployed drift anchor, so it was safe for both of us to get in at the same time. I swum away from the boat. Normally we only went over the side to work on cleaning the bottom, so it was great to have a leisurely swim. Once I got about 10 metres away I turned and looked at *Yantu*. She looked beautiful against the clear blue sky and water. Were it not for a few sun-faded logos on the side and algae growth under the scupper we used for pouring out the dirty dish water, she looked as if it was her first day in the water. Sun Haibin joined me and we then had a race around the boat, winner Sun Haibin, before we got back on board to rinse off the salt water and check the fishing line. We had not caught anything, not even a jellyfish. I suggested to Sun Haibin that we leave the line out, but he did not want to do so. While fooling around in the water he had reflected on the concept of fishing and had decided that killing something on the ocean when we were ourselves living at the mercy of it would be tempting fate and result in bad luck for ourselves. I found this a bit strange, but was happy to respect his view. But I was more coming at it from the angle that, if we caught something, we would most likely have to stop to pull it in and prepare it. That would slow us down and we needed to get to Barbados!

As we got closer to Barbados, we started to experience the effect of Hurricane Olga. The hurricane season was already over, but Olga was a late bloomer and not sure where she was heading. She was to the north, but then headed south towards us before going north again. She brought with her falling pressure, lower temperatures, big winds, big waves, thunder and lightning and we had no desire to meet with her, particularly not with compromised hatches. She was certainly influencing our weather pattern. It was like she was stirring a big pot and we were sitting in it.

> *Log entry — Day 50 — 26th November*
> Wind force 2-3 SW, later 2-3 SE. Waves 1-2 metres. Barometer 1006. Cloudy and rain, 27°C.
> Started rowing again at 5:30AM as wind shifted to SE.

It rained a lot. I liked it, but Sun Haibin did not enjoy getting wet and cold. On that day he rowed his watch in the rain, then the rain cloud went away and I rowed my watch in dry weather. As my watch was finishing, I could see the cloud come back. It started dripping and as Sun Haibin started rowing, it rained on him and continued for another two hours straight! It was quite a funny coincidence, but it left Sun Haibin in a gloomy mood.

Sun Haibin rowing in the rain and big waves.

'I can't believe the weather is turning on me like this. It is not fair!' he said and disappeared into the cabin to dry himself.

We had slowly been closing in on *Atlantic Warrior* for the past week and they were now our target to overtake. They were less than 20 nautical miles in front of us.

> *Log entry — Day 51 — 28th November*
> Wind force 0-1 NE, night 6-7 N for three hours. Waves 1-3 metres. Barometer 1006. Sunny, then rain, 26°C.
> Rowed all night and did 5 nautical miles per shift! Great after a very hot day where we only did 3 nautical miles per shift. Total distance covered was 50 nautical miles.
> Rowed at 8AM this morning when a big cloud appeared with rain and headwinds. Put out drift anchor, but only needed for 20 minutes. Cloud cleared and progress now good.

The last thousand miles

> Rowed four hours through rain and waves at night. Quite bad. Put on life-jacket and lifeline. Was thrown out of my seat by big wave.
> We passed *Atlantic Warrior*. Now in 8th position!

I was rowing the 9–11PM shift when the clouds got thicker and uglier and the rain intensified, so I asked Sun Haibin to hand me my life-jacket and lifeline. As I was putting it on the weather deteriorated further and the wind started blowing from the north. I clipped on my lifeline to the windward jackstay and continued rowing. The wind kept increasing and the waves started building. Within half an hour they were three to four metres high and getting steeper. They started breaking over the side of the boat as the wind intensified. Luckily it then started raining hard. This helped flatten the sea and stopped the waves from breaking. It was cold and miserable, but at the same time invigorating. I felt very much alive.

This is just a cold front. I can row through this. If we don't stop I am sure we will have passed Atlantic Warrior *tomorrow*, I said to myself, and continued rowing.

Because the weather was really bad and it was raining hard, I decided not to wake up Sun Haibin. This was my opportunity to repay him for rowing my shift a few days earlier. I kept going and eventually the rain stopped and the wind died away. We had passed through the front. I rowed on, enjoying myself. The waves were now back to a reasonable size so I did not pay too much attention to them, but suddenly a freak wave threw *Yantu* onto her side. I fell out of my rowing seat and landed at the end of my lifeline against the lower stanchion. My feet were still attached in my shoes and as I was struggling to get back in my seat, Sun Haibin opened the hatch.

'Sounded like really bad weather. What happened?' he asked.

'Oh, freak wave. No worries,' I replied. 'Good you came out. I wouldn't mind a rest.'

Sun Haibin looked at his watch. 'But you have half an hour left.'

'Look again.'

'What, you rowed for four-and-a-half hours in this weather! You're crazy!'

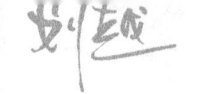

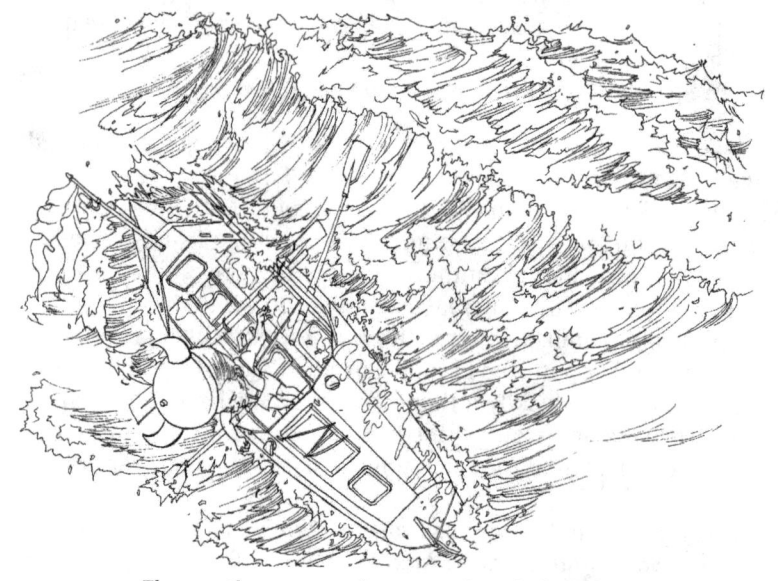

The wave threw me out of my seat to the end of my lifeline.

'I did not want to disturb you because it was raining, but would you mind taking over now?' I replied.

Sun Haibin came out and cooked soup noodles for our night snack. We looked at the dark cloud cover behind us. He then started rowing and I went into the cabin to rest. The cabin was wet. The rain and waves had made it through the cabin roof hatch.

I tried to sleep, but my mind was racing. It had been a bit unsettling to row through the rainstorm with the compromised cabin roof hatch. *It would be nice to get a visit from one of the rescue vessels,* I thought before I finally fell asleep.

The Challenge Business must have heard me, because later in the night one of the safety vessels arrived. We talked over the VHF for a while and they told us that they would stay close to us until it was light. We did not want this out of fear of breaking our rhythm, so we asked them to sail away and come back when it was light. They came back the following morning and we talked some more. They told us the Belgian brothers had arrived in Barbados looking like concentration camp victims. The crew on the safety vessel were the first people we had seen since leaving Tenerife. It was strange, but also a sign that the race was almost over. After a

The last thousand miles

while they motored off, but then they seemed to stop again. I got out the VHF.

'Do you guys need a tow or something?' I teased them.

'No, we are deciding who to visit next. Maybe *Team Manpower*, they are making very good progress.'

They disappeared in the still big waves. We saw them again after we arrived in Barbados, by which time they had lost their mast due to the big rolling waves.

Visit by the race safety vessel after 52 days at sea. The next time we saw the safety vessel it had been demasted thanks to Tropical Hurricane Olga, which also gave us a good licking.

Log entry — Day 52 — 29th November
Wind force 1-2 E. Waves 1-2 metres. Barometer 1007. Sunny, 27°C.
Excellent rowing conditions, but only did 49 nautical miles.
No rain — good going.
We are feeling tired, but in good spirits.

During the night we had seen a very strange ship, judging from its lights. It looked like it was sailing sideways. Intrigued, we decided to call it up on the VHF radio.

'*Yantu* calling lit-up ship. Come in, over,' I called.

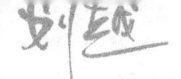

'This is the *Wind Surf,* over,' a distinctly upper-class British voice announced.

'What kind of a ship are you? You look like a lit-up Christmas tree, over.'

'We are the world's largest cruising sail ship. We carry 300 passengers and 150 crew, over,' the voice announced proudly.

'Then I hope you don't have to evacuate ship onto us. We are a seven metre long rowing boat with a crew of two,' I replied to see if he had any sense of humour. No reply. I continued:

'Is there any chance you can see another boat like ours on your radar?'

'You must mean *Atlantic Warrior*. We spoke to them a few hours ago. Just a moment.' The voice went away and came back. 'You are 14 nautical miles ahead. Anything else I can do?'

'No, that is great. Thank you. Have a good journey. Out.'

I turned to Sun Haibin. 'Excellent news. We are 14 nautical miles ahead of *Atlantic Warrior*. I bet they stopped rowing during the rain storm!'

> *Log entry — Day 54 — 1st December*
> Wind force 4-5 E. Waves 3-5 metres. Barometer 1008. Sunny, 29°C.
> Excellent rowing conditions. Did our best distance to date — 62 nautical miles and already making excellent progress on tomorrow's 52 nautical miles. Thank God for a good wind to finish with. The log estimates 32 hours left to Barbados!
> Great feeling to know this is the last 100 nautical miles!
> We are in good spirits and quite excited because the end is so near!
> Called Véronique last night in Barbados. Also spoke to my sister and mother. Great they are all there!
> *Atlantic Warrior* 20 nautical miles behind us this morning.
> My butt is bleeding into the lambskin and it sticks to me when I stand up, but not much pain. The painkillers seem to be working.
> We will get there for Véronique's birthday!

That was my last log entry. We were on a complete high as we continued rowing on 2nd December. We were in eighth position, well ahead of *Atlantic Warrior* and with the favourable wind and waves we were confident that we would make it for Véronique's birthday on 3rd December.

After so much effort and pain we were going to deliver on our promise — finish in the top 10 and in 56 days! It felt *great!* And even better, we were in eighth position. It seemed fitting that lucky boat number 18 should finish in lucky number 8 position! It was a good Chinese omen!

Sprint to the finish

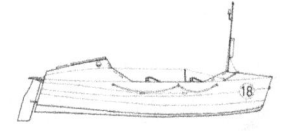

Sun haibin was rowing early morning on 3rd December and I was resting in the cabin.

'Christian, come out here!'

I rushed out of the cabin, not knowing what to expect. 'What's up?'

'Look,' said Sun Haibin excited and pointed to a grey blob on the horizon. 'It's *Barbados!*'

I looked at the grey blob for a while. It stayed stationary on the horizon and it seemed to have trees on it. He was right. It was Barbados!

I got out the video camera and filmed Sun Haibin rowing towards Barbados, his face beaming in a broad smile. We joked around for a while and decided to have breakfast. Sun Haibin then went in to rest and I rowed my shift.

Barbados on the horizon. First time we see land in 56 days. Sun Haibin's smile says it all!

Sprint to the finish

As Barbados got closer and clearer I started feeling strange. I had actually done it. My crazy project had succeeded, but instead of making me happy, it brought a lot of questions. What are you going to do now? For the past year and a half I had never thought about life after the race. Right now I was an adventurer, but in about 10 hours we would have finished the race and I would then be just a person out of a job. What I really wanted was a rest. I did not feel I had the energy to start the whole new project of getting back into normal life. Suddenly my urge to finish the race disappeared and along with that my ability to row. Sun Haibin came back out.

'What's the matter with you?' he asked.

'I don't really want the race to finish,' I replied.

'You must be joking!'

I told him how I was feeling. He told me he was feeling just the opposite. He was pretty sure he would be a hero in China and that the university would help him pass his exams and give him a job. Sounded great!

Since I was useless on the oars, Sun Haibin rowed the rest of my shift while we talked and slowly my mood came back.

The second Challenge Business safety vessel came out to visit us. Because of the heavy surf on the rocky eastern coast of Barbados we were going to be escorted in by the safety vessel, just in case we got into trouble. As mentioned before, ocean rowing boats are in the most danger when close to shore.

The safety vessel came by and we talked on the VHF.

'You guys better hurry up if you don't want to get beaten by the boat behind you,' they said.

We knew we were 14 nautical miles ahead of *Atlantic Warrior* and with only twenty miles to go there was no way they were going to catch us up, so we did not pay attention to the warning.

Instead we decided to take a shower and shave so that we would arrive in Barbados looking decent. I was standing in the cockpit shaving when I looked south. What the hell! Another rowing boat! I could not believe it. I yelled to Sun Haibin who was working up a nice lather in his hair with the shampoo. He also spotted the boat. It was *Team Manpower*, coming out of nowhere, and they were less than 300 metres away!

We did not stop to marvel at the fact that after racing 2,700 nautical miles across the Atlantic, it was possible for two boats to arrive on the other side and be only 300 metres apart.

We jumped back on the oars and rowed together. All my lethargy was gone. The speed shot up to five knots. There was no way we were going to be overtaken in the last few hours of the race! And there was no way we were going to give up our lucky number eight position!

On *Team Manpower* Richard White and Ian Roots had other plans. For the past few days they had gone flat out to overtake us. They had thrown weight overboard, stopped sleeping and had managed to cover 80 nautical miles per day. A fantastic speed! They wanted to make their grand finale by beating *Yantu*.

The race to the finish was on!

Race rules stipulated that we had to go round the north of Barbados. The race finish was at Port St Charles, a marina pretty much halfway up the west coast. We were now halfway up the east coast of Barbados so we still needed to get around the island.

Since we were 300 metres north of *Team Manpower* we were in the lead. We rowed until it started to become dark and kept the distance.

I dropped out to ask the safety vessels over the VHF for recommended waypoints to get safely around the island. I got back on the oars. Still 300 metres in the lead.

Team Manpower was less than 300 metres away.

Sprint to the finish

The safety vessel came back on the VHF. I dropped out again to code the waypoints into the GPS and then got back on my rowing station with the GPS next to me. The gap had narrowed significantly and we were now abreast, with *Yantu* being further out to sea.

We followed the waypoint round the top of Barbados, called North Point, and on the western side we continued rowing out to sea to get to the allotted waypoint. Suddenly we were so far off land that we left the wind shadow of the island. The westerly wind was now pushing us offshore and as we turned the waypoint to come back in, the wind was on the nose. For a split second we worried that we might not be able to row back. *Team Manpower* was much further in towards land, and the safety vessel was following close behind it.

What was going on? Had we been given dud waypoints? It looked like we had been sent on a losing course. We were not happy. We rowed on and came back into the wind shadow. We were still level.

It was getting dark so we switched on the navigation light. Soon all we could see of *Team Manpower* was their navigation light close by. We were still level.

'Come on, let's go for it. These guys are not going to beat us! Let's row them into the ground!' I said to Sun Haibin.

We doubled our efforts and slowly we started pulling away. One hundred metres, two hundred metres. Suddenly we sensed that *Team Manpower* had given up the fight. We were now pulling away more easily. A speedboat with supporters came out of the dark and its big wake hit us and stopped us dead. It circled back and we got hit by its wake again. It drew level with us.

An excited voice yelled down to us: 'You guys are doing five knots.'

I yelled back: 'Get that wake out of our way NOW. You are ruining our stroke!'

Beijing to Barbados in a Rowboat

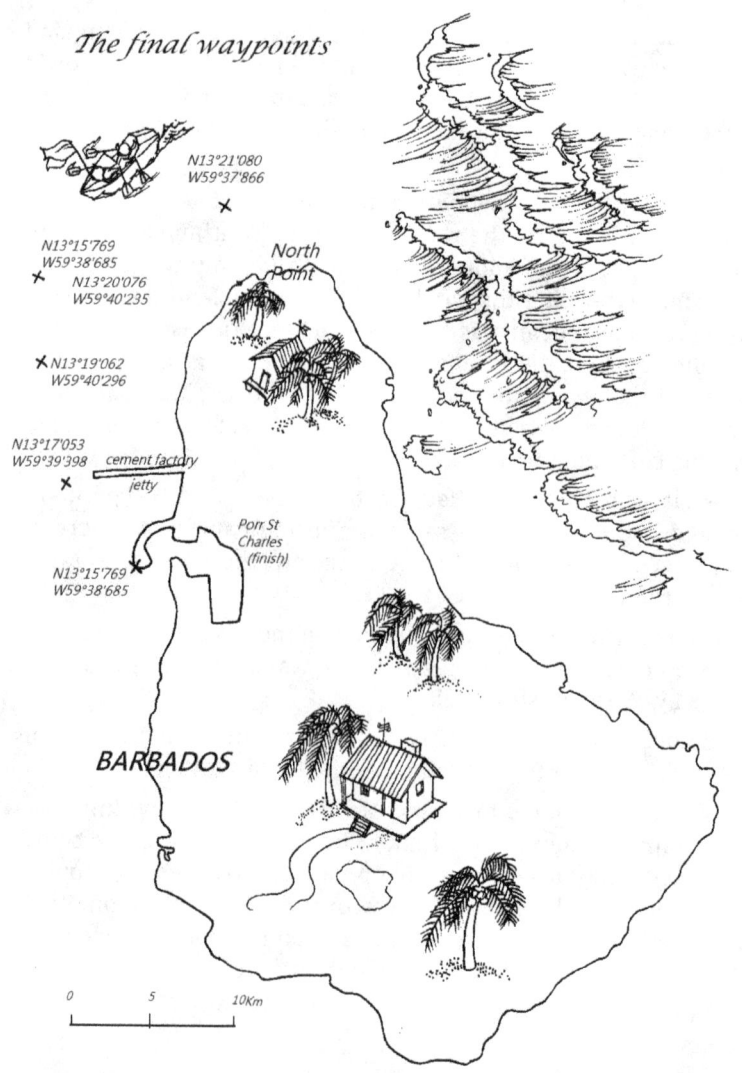

The final waypoints

N13°21'080
W59°37'866

N13°15'769
W59°38'685

N13°20'076
W59°40'235

North Point

N13°19'062
W59°40'296

N13°17'053
W59°39'398 — cement factory / jetty

Porr St Charles (finish)

N13°15'769
W59°38'685

BARBADOS

0   5   10Km

Barbados map waypoints.

The speedboat got the point and sped off to say hello to *Team Manpower*.

We were still ahead and *Team Manpower*'s navigation light got further and further away.

'We are winning! We are winning!' Sun Haibin yelled with joy. 'When we get to Barbados I am going to hug you!' This is a very unusual thing

Sprint to the finish

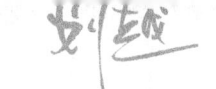

for a Chinese to say, and it drove home the feeling of comradeship and excitement we both felt.

We rowed on. As we got closer to the finish we suddenly smelled wet soil. We were arriving. We then smelled the perfume of the supporters who were standing on the breakwater looking at the battle between the two small white navigation lights, trying to guess who would arrive first.

'Sun Haibin, I need to stop rowing,' I suddenly said.

'Why now?'

'I am not wearing any clothes! I can't arrive naked!'

I climbed out of my rowing seat and put on some shorts. It felt strange to wear shorts again after thirty-odd days of being naked. I got back on my rowing seat. The breakwater was now so close that we could see the mob of people standing on it.

As we came closer the noise level increased. People were yelling and screaming encouraging us on.

Suddenly someone on the breakwater yelled: 'It's *Yantu!*' There was a wild cheer. We zoomed past the breakwater.

Not sure where the finishing line was I yelled: 'Have we finished yet? Have we finished yet?'

The answer was drowned out by the crowd. We had finished! The time was 21:07 Barbados time and 01:07 GMT. Our finishing time was 56 days, 15 hours and 52 minutes. Fifty-six days and in the top 10! Mission accomplished! It felt great and we both yelled our heads off for a while. We then rowed to the jetty to disembark.

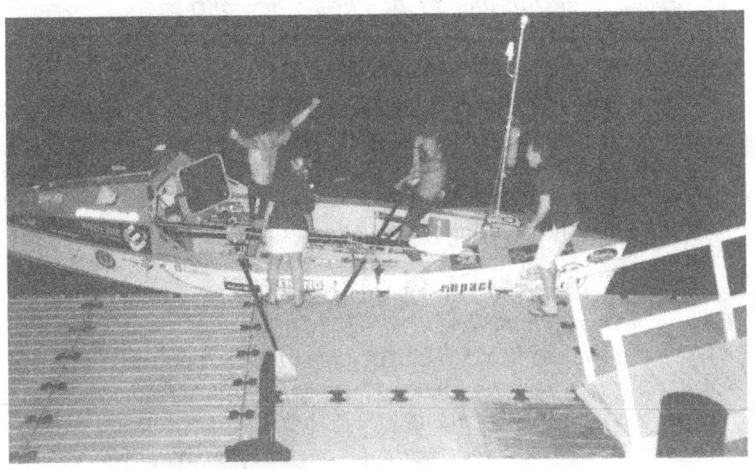

Finished! All objectives achieved and still friends!

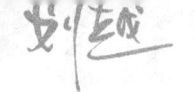

Teresa Evans from the Challenge Business was standing on the jetty to greet us. Above her on the pier the supporters were looking down at us. They were not allowed onto the pier since we had not yet cleared customs. I tried to make out Véronique in the crowd, but I could not. It was too dark and there were too many people. I was dying to see her, but we still had work to do. We slowly tidied up the boat, got our passports out and got ready to step onto solid ground for the first time in 56 days.

The first step on land was very strange. I almost fell over. We were not used to standing on a non-moving surface. We hugged and stretched our hands into the air towards the supporters above. We then walked, or rather staggered, into the customs building to clear customs and immigration. Sun Haibin's visa for Barbados ran out 3rd December, so we had arrived with just three hours to spare!

We had plastic chairs to sit on while filling out the papers as well as a *cold* Coca-Cola. Unbelievable luxury! But we had problems handling the pen to fill in the papers. Our fingers would not grip the pen, but eventually we were done. It was time to go outside to meet Véronique, my mother, my sister and all the other supporters.

It was pure joy to see them. I hugged and kissed Véronique and wished her a happy birthday. Sun Haibin yelled 'Mom,' and then hugged my mother as if he had known her all his life. My sister also got a hug.

Our arrival had also been very tense for the supporters on land due to the close finish with *Team Manpower*. Véronique's diary reads:

We drove to North Point to look for Yantu. It was a beautiful spot, beautiful cliffs and a magnificent view of the sea. We arrived for sunset, which was beautiful, and the wind in our faces gave us the feeling of grand nature.

We first spotted the safety vessel with its high sail, then when it became dark we could see your navigation light very well. It was the first time I saw Yantu by night. Even though you had good speed you looked so slow! But we stayed long enough to see you move. It was great to 'see you', or rather, know it was you out there. At the same time, I was worried because this coastal approach was the most dangerous part of the trip, and also because of Team Manpower.

We drove back to Port St Charles and stopped at a nice restaurant for a rum cocktail to make time go by a little faster. From there I made a call to the office to check on the situation and the latest ETA. It was 9 to 9:30PM. That was fine. What was not fine at all though, was the fact that Team Manpower was half a

Sprint to the finish

mile behind you. Half a mile! I asked whether you were aware of that and the response was: 'They are so close they can see each other.' The piece of good news was that you had cleared North Point and were now really completely safe.

We drove down to Port St Charles a little later on, and I was obsessed with the thought that instead of truly enjoying the last five miles like you both deserved, you were really suffering, physically because you had to sprint, and mentally because you might be very angry if you had been or were being overtaken.

When we arrived at Port St Charles there were lots of people. Everyone was excited, but of course we were nervous. Teresa had told me that they thought you had been overtaken. A boat that had gone to see you confirmed there was a half-hour difference between the boats, but did not confirm the order. The first boat was obviously coming so fast and I resigned myself to you coming 9th. Then, shortly before the end, it appeared Teresa did not know for sure who was first. We could only start (but barely) seeing this rowing boat approaching when it was 100 metres away. I looked with my binoculars and thought I could recognise Yantu with its colours, its logos, and the quite high mast. But since I didn't know what the other boat looked like (potentially very similar), I did not dare to hope....

I don't know when it became absolutely clear that it was you, and who had that certitude first, but once it was clear, I was really ecstatic! This last, very close and nerve-wrecking run-up made me even happier and prouder to see you arrive. I was yelling like a mad woman!

The sight of you both bare-chested, rowing elegantly and swiftly in the dark, was really dramatic. You really looked like real adventurers! I could also see that you were quite tanned and had lost weight.

You took forever to put the oars back into place and step out of the boat (your first attempt to get up ended up making us all laugh).

Then we were waiting. I felt almost numb, nervous, but not really happy or relieved quite yet; there had been so much tension. Finally you came out and we had the first kiss and the first eye contract since Tenerife.

* * * *

We hung around and spoke to the other supporters and rowers who had already arrived. After a while they all disappeared out to the breakwater again to welcome *Team Manpower*. They arrived exactly 24 minutes after us. A close race or what!

We talked a bit with the *Team Manpower* team and then it was time to call it a day. We were tired. We piled into Véronique's rental car and headed for our hotel. Véronique wrote:

When we came back to the hotel, Sun Haibin shouted in delight at the sight of the bed and the size of the room and it made me laugh. We all spent time chatting in his room until we were too tired. It was great to see both of you still laughing so much!

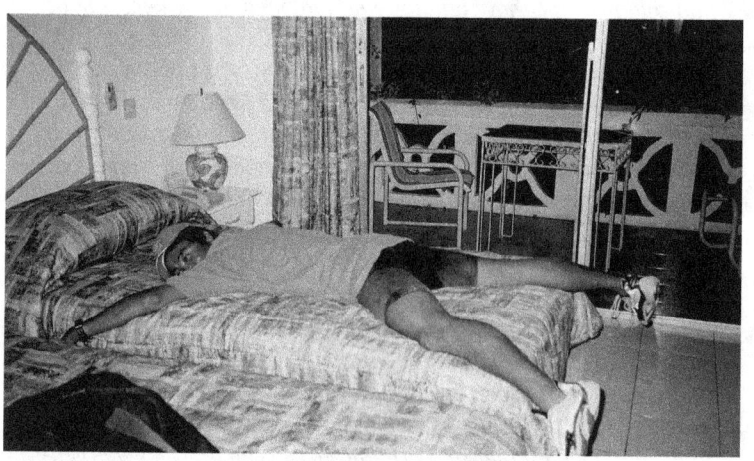

Sun Haibin savouring being back on terra firma in a decent bed.

I slept well that night and not once did Sun Haibin call me to get on watch!

Hanging out in Barbados

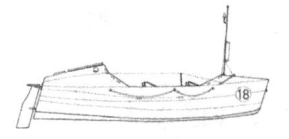

The next days were days of celebration. I would get up around nine and the whole room would be moving around as if it was floating on a big wave. I would almost get seasick, but when I got out of the shower the sensation had stopped. We would then go to breakfast with a funny walk. However hard we tried we could not walk straight and we looked like a pair of drunken sailors. It took four days for our bodies to readjust to land.

Breakfast at the hotel was excellent. It was a buffet with fresh fruit, pancakes, yoghurt, bacon, eggs — you name it, it was there — and we ate it all. But not before we had asked for extra cushions to sit on! Our butts were still very sore and remained so for a long time. I wore a sarong because the skin was too raw for me to wear shorts. We had problems holding the knives and forks as well. Our fingers had been shaped around the oar handle and now stayed like that so that we were absolutely unable to close our fists, particularly since the joints were no longer exercised every two hours. We looked like Captain Hook! It took Sun Haibin a week to get his fingers back to normal and me about a month. A few months later I eventually went to see a doctor about my ingrown toenail and then that healed as well.

We re-celebrated Sun Haibin's birthday. He got a picture frame with a picture of both of us at the start in Tenerife from Véronique, a plate with the Little Mermaid motif on it from my mother and a Christmas ball from my sister, one of those you turn upside down and shake and then turn around again to watch the snow falling.

'I thought you might enjoy the cold,' my sister said.

Sun Haibin celebrating his birthday with my mother and sister.

Sun Haibin was ecstatic about all the presents and the champagne made our moods even better. Véronique went to send out the press release she had written and later came back with literally hundreds of congratulatory e-mails. It was amazing. We sat down to read them. They were great and we almost felt embarrassed! I especially appreciated those from friends in Denmark; Duthie at the Royal Hong Kong Yacht Club; Malcolm McKenzie, principal of Atlantic College and from Chris Perry, who had tried to teach us how to row:

> I can only express my absolute admiration at their truly remarkable achievement. Not only have they achieved their objective of finishing the race, but they have done so in flying style by finishing 8th overall.
> Despite my rather derogatory comments about their rowing technique when I went out with them on the harbour in Hong Kong, they will be welcome any time at the Hong Kong Team boathouse!

Sun Haibin called his girlfriend Cao Xinxin in Beijing. She told him that more than one hundred journalists had called her from all around the country and that she eventually had had to switch off her mobile. It sounded like Sun Haibin was indeed going to return home to China as a hero.

Hanging out in Barbados

The following evening we celebrated Véronique's birthday at a restaurant. We were sitting outside (on cushions) and enjoying the evening, when suddenly Sun Haibin and I started. We had both heard the WUUUUUSH sound of a breaking wave. We tensed and looked around. The noise was coming from a passing car with something wrapped in plastic on the roof rack.

We could not help but laugh at ourselves. Véronique, my mother and my sister looked at us not knowing what was going on.

* * * *

A highlight of our stay in Barbados was to go down on the breakwater and greet the other teams as they were arriving.

Atlantic Warrior arrived one day after us and the Americans Tom and John, rowing *American Star*, the next day. The Americans' arrival was very emotional for us. We knew the guys well from Tenerife and we had been battling to beat them. They, and their support crew, were the people we knew the best and we were yelling like mad from the breakwater to encourage them on. Tom's girlfriend Sarah was beside herself with joy and relief. Véronique and Sarah had been exchanging e-mails and, like Véronique, she had worried about her guys out at sea. One of Sarah's e-mails to Véronique reads:

Tom and John are doing OK, they are tired and just want to get to Barbados. Tom hurt his knee, and you know about John's crotch rash (I can't believe Tom told him to try and pour vodka on it when they have a whole first aid kit with them! MEN!) I talk to Tom almost every day, and sometimes it is sooo hard. I get off the phone and want to cry and scream because there's nothing I can do to help and it can be such a drain. I feel like I'm on a roller coaster, and I hate roller coasters. But it will end and all I want is for our guys to land safe and sound.

We were all yelling with tears in our eyes. Tom looked somewhat the worse for wear, but recovered fine over the next few days. When they came out we shook hands. Sun Haibin was very pleased to see Tom again.

Tom shook his finger at us: 'You guys, you guys ... I can't believe you overtook us.'

We hung around on the breakwater for the next hour absorbing the pure joy around us. In a way Tom and John's finish was also our finish. We had looked forward to rowing across the finishing line at a leisurely pace while reflecting on the race and our achievement, but instead we had sprinted like crazy and had had no time to think about anything. Seeing the Americans arrive enabled us to experience the finish we had looked forward to.

My sister and mother went back to Denmark and for the next days Sun Haibin, Véronique and I hung out on the beach and breakwater. One day a couple, Kevin and Sandy Reath, came across to talk to us. They were living at Port St Charles Marina and had seen that *Yantu* had a Royal Hong Kong Yacht Club logo on her side. They used to be members there and would therefore like to look after us now that we were here in Barbados. What a kind offer! We gladly accepted. Over the next days we were treated to lobster and went fishing from their speedboat. Cruising along at 30 knots, beer in hand, was something quite different from rowing *Yantu* and Sun Haibin was excited when he had a go at helming.

Another readjustment we had to make was to get used to speed on land. Although the coastal road in Barbados seldom allowed us to drive faster than 40 km per hour it felt like much more.

Véronique left Barbados and Sun Haibin and I were alone again. We went about cleaning up *Yantu*, got her out of the water and arranged for her to be shipped to Denmark where she would be exhibited at the Copenhagen Boat Show. Our butts were recovering and we could begin to entertain the idea of sitting on a plane seat for twenty-odd hours to Hong Kong.

A few days before we flew out Port St Charles hosted a party for the rowers and supporters. I went around with my video camera asking all the rowers the same three questions: 'Why did you want to row the Atlantic', 'Did rowing the Atlantic change you?' and 'Would you do it again?' I got several colourful replies.

Like myself, the other rowers also found it difficult to articulate what had made them want to row the Atlantic:

- 'It is the first idea I have seen that has grabbed me instantly. It is like being attracted to someone. You don't know you will be attracted to that person until your paths cross and then it just happens.'
- 'To get to Barbados on the other side.'
- 'I did not want to sit behind a desk until age 65 doing masses of overtime at work.'
- 'God knows — too many beers. Never decide to do something when you have had too many beers.'
- 'My partner promised me he would get me drunk and laid if I signed up, so I did it.'

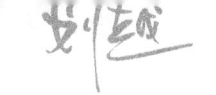

- 'That is the million dollar question — I have been asked it loads of times and I really don't know.'
- 'Fate!'

As to whether rowing across the Atlantic had changed them there was a wider range of replies ranging from the obvious:

- 'It changed the shape of my ass. My ass will never be the same again!'
- 'I lost a couple of stone.'
- 'I have a knee that does not work too well any more and a rash on my chest that will leave me permanently scarred.'

... to the more profound:

- 'I don't know. I don't think I have, but maybe other people will see changes in me when I get back. The challenge was 80 percent mental and 20 percent physical.'
- 'Yes! I learned a lot about myself and about how people develop in teams. It has been really exciting.'
- 'I feel more alive in the world. I will enjoy everything more. I will have more interest in things I have never thought about before and the small things that used to bother me — they are not important any more.'

... and the more reflective:

- 'Until seven days before the end I thought I had learned a lot about patience ... I thought I could live a bit better with the fact that you just have to let some things happen and in the end it will be OK. There are things you can't control ... but then the wind changed ... suddenly I realised I was very bad at dealing with patience!'

As to the question of whether rowing the Atlantic was a worthwhile thing to do again, there were a few positive answers:

- 'Right now I say never again, but ask me in a month's time when my body has healed and I might have changed my mind.'
- 'I am telling everyone "no" at the moment, but actually I probably would.'

But the majority had had enough of ocean rowing, if not of adventure:

- 'Not row it. The idea of spending two years consumed by preparation to do the same event again? No way! It takes too much out of you and everyone around you. It would have to be a new adventure to make it worthwhile. I might sail the Atlantic.'

- 'It is not called the world's toughest rowing race for nothing. It was incredibly long. A lot of lows and a few highs. I might do something else, though, like climb Mount Everest.'
- 'Never, ever, ever! I've got better things to do than to spend two months in a rowing boat. I have done it and I enjoyed it. Next time I will do something different like driving down South America or motorbike across Africa. Whatever it is, it will be on *terra firma!*'
- 'Life is too short not to row an ocean and too short to do it twice.'

Finally, Pedro Ripol's replies from the Tenerife entry *Martha 2*, hit home:

'Why did you want to row the Atlantic?'
> 'That is a question I keep asking myself. I think you have to have an adventurous spirit. It is a challenge. You get to know yourself better. You appreciate a lot more small things like your family, your friends ... you take care to enjoy them, to be with them. You give your energy to the people around you and it comes back to you multiplied by thousands.'

'Did rowing the Atlantic change you?'
> 'I think rowing an ocean changes everyone. Even if people consciously don't think so, I think it does. Forty-five to 90 days rowing 10, 12, 16 hours a day in the middle of nowhere makes you think a lot ... it makes you appreciate life a lot. The majority of people spend their lives watching TV. That is not my way.'

'Would you ever row the Atlantic again?'
> 'For two million dollars — no problem!'

'How about for two dollars?'
> 'If it is with Brook Shields — maybe. Depends on who my next rowing partner is. If I get to sleep in the cabin with a beautiful girl I would not mind, but with my partner Pancho...? we enjoyed it this time, but enough is enough!'

* * * *

Hanging out in Barbados

Shortly before leaving Tom asked Sun Haibin and me to come to his hotel room as he had a present for Sun Haibin. When we arrived Tom was in a sober mood and told me to translate.

'Sun Haibin, we flew this Stars and Stripes flag every day at sea. It is hand stitched and difficult to come by. I want you to have it as a gesture of our friendship as well as friendship between China and America.' I felt like a younger version of Henry Kissinger as I translated. Sun Haibin was very moved as he accepted the flag. It was a fine way for them to consolidate their friendship.

Sun Haibin and Tom to Barbados. The flag Tom presented to Sun Haibin as a gesture of Sino-American friendship.

We were done in Barbados. It was time to go back home. On 11th December we boarded a plane heading back across the Atlantic. Reclining in our seats and with a drink in hand, we were looking at the map on the flight monitor. After only a few hours of flying I pointed at the monitor and said to Sun Haibin: 'This is where we were on Day 30.' We both laughed.

Back to China

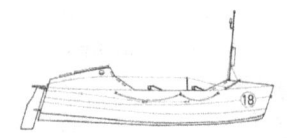

We changed planes in London and flew to Hong Kong on Cathay Pacific, who kindly upgraded Asia's first ocean rowers to business class. We arrived back in Hong Kong on 13th December 2001. Because Sun Haibin had an onward ticket to Beijing, he was allowed to stay in Hong Kong for three days.

In the evening we went down to the Royal Hong Kong Yacht Club who had put on a celebration party for us. There were more than a hundred people crammed into the Chart Room and the beer was flowing.

It suddenly struck us that what we had done was perhaps a bit out of the ordinary. In Barbados we had not been thinking of this since most people we talked to there had also just rowed across the Atlantic. It had seemed pretty ordinary and nothing special. Seeing the people there to greet us, and feeling the atmosphere of the room, made us realise that there was a good reason to celebrate.

Duthie came across the room and shoved a beer into our hands: 'You guys did well. I can't believe it. You actually look as if you enjoyed it!'

The rest of the evening was a blur. The next morning we had a press conference at the Yacht Club with plenty of Chinese and English TV camera crews as well as journalists.

We were also invited as honourable guests to the Royal Hong Kong Yacht Club Rowing Section's Annual Christmas Party where we had more beer. During the party we found out that the Rowing Section had entered us into their rowing ladder on a handicap basis at the start of the race and that we were currently in third position. We could expect to receive challenges for the position, we were told. 'I hope it is not in a racing scull. I can't row a thing like that,' I said. Bad luck, it was. I suggested I would only take challenges over distances of more than 1,000 km, but that idea was vetoed. When I later started taking up the challenges on behalf of Sun Haibin and myself I had several capsizes and *Yantu* quickly slipped down the list. Racing sculls just weren't for us!

Welcome back banner at the Royal Hong Kong Yacht Club and press conference.

We had more good news. The Yacht Club had received an additional US$35,000 towards scholarships. It looked like we would be able to send another student to Atlantic College! What a great feeling!

On 16th December we flew to Beijing. I was teasing Sun Haibin all the way that he was now a big hero, but we were taken aback by the welcoming committee in the airport.

A hero's welcome at Beijing's Airport. The yellow banner reads: 'Triumphant Return for classmate Sun Haibin.' We also starred on CCTV5's exclusive "Five Rings Night Show", a show normally reserved for professional Olympians and World Champions.

The Beijing Sports University was out in force along with numerous television crews and journalists. It was overwhelming. We could see the 'welcome home' banners even before we came out through customs. Zhang Jian was there and Sun Haibin was reunited with his girlfriend Cao Xinxin. We were then bussed off for a banquet while the press kept interviewing us.

We made the front page of Beijing Youth Daily, Beijing's most popular newspaper with a daily circulation of 660,000 copies, for three days as well as several other papers. We were on the news and the phone did not stop ringing. We did a talk show and then a documentary. It was a flurry. The Beijing Sports University decided to reimburse Sun Haibin his tuition fees.

We were asked to give a talk in front of the students and teachers at Beijing Sports University. That night we found ourselves up on a podium looking out on a full lecture hall with close to 1,000 students and teachers.

This was Sun Haibin's show and I let him do all the talking. He was so excited that he was unable to say anything sensible, but it did not matter. The excitement in the room carried him through.

Christmas Eve was getting close and I left Sun Haibin to enjoy his new found fame. I had promised Véronique I would celebrate Christmas with her at her parents' house in the French Alps. I looked forward to relaxing and recovering. 'What better place to recover than with Véronique in front of a log fire and with snow outside?' I said to myself as I made my way to Beijing Airport. My adventure was over and as the plane took off I fell into a deep and contented sleep.

We were featured three days on the front page of Beijing's most popular newspaper, Beijing Youth Daily, with a daily distribution of 666,000 copies.

Postscript

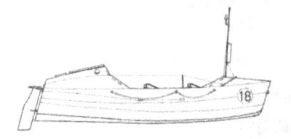

Yantu was shipped from Barbados to Copenhagen where she was exhibited at the Copenhagen Boat show in February 2002. With the assistance of the United World Colleges Network in Denmark we raised some additional scholarship funds for Mainland Chinese students during the show and His Royal Highness Crown Prince Fredrik, who is patron of United World Colleges in Denmark, visited our stand.

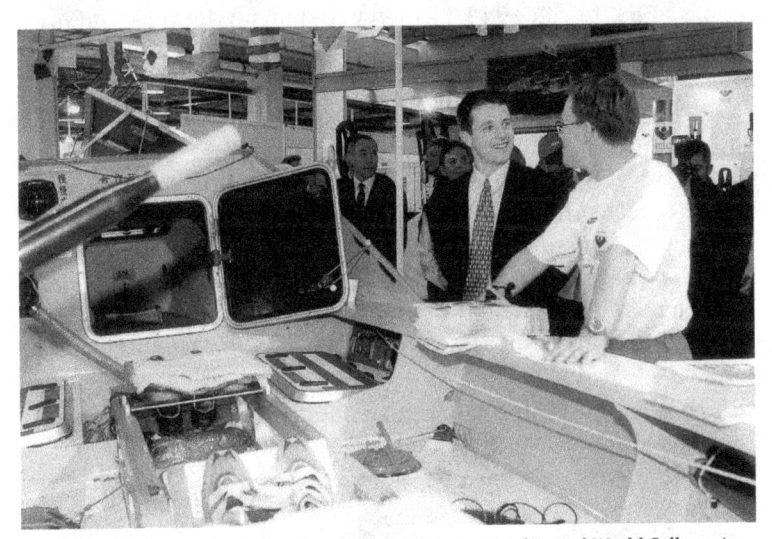

HRH Crown Prince Frederik of Denmark and Protector of United World Colleges in Denmark, visited our stand during the 2002 Copenhagen Boat Show. Having trained as a Navy Seal, we had not chatted long before he looked me straight in the eye, and exclaimed: "Your ass must have hurt like hell!". What a top guy!

Through the efforts of UWC alumna Michael Yong-Haron in Hong Kong we managed to raise the remaining funds needed to send the second Chinese student to Atlantic College. Hui Wang started studying at Atlantic College in September 2002.

I later sold *Yantu* to Richard Pullan and Chris Hall, competitors in the 2003 Atlantic Rowing race, in order to recover some race costs. They also had an education charity aspect to their entry and I felt *Yantu* passed on to good hands. They even kept lucky race number 18, and secured five-time Olympic rowing gold-medallist Steve Redgrave as their patron.

During the year after the race I did some consulting work in Taiwan and was then asked by Denmark's best-known travel writer, Troels Kløvedal, to sail with him to China to make documentaries for Danish television. I had originally met him in Hong Kong, as I was preparing to send *Yantu* to Tenerife, when he had been trying to get the right permits to sail to China. I had given him some ideas and contacts and he subsequently managed to obtain the most extensive cruising permit for China since the founding of the People's Republic in 1949. I sailed with him up the coast of China for about a month, but I was suffering from adventure overdose so I left to go back to Véronique in Munich, where I sat around in a sunny park for a while pondering what to do next.

Late 2002 Véronique and I got married and she got pregnant. We moved to Shanghai where I ended up in a job that requires me to wear a suit again! However, after several hours of therapy I now appear to be adjusting reasonably well.... Our son Victor was born in Shanghai in May 2003 and thereby followed the modern commercial trend of being 'engineered in Germany and manufactured in China', like most other quality products nowadays!

In connection with our Chinese book launch Sun Haibin and I organised China's first amateur indoor rowing competition at the 7th Shanghai International Boat Show together with Concept II.

China is an exciting place. There are so many opportunities and adventures waiting to be undertaken, so watch this space.

Postscript

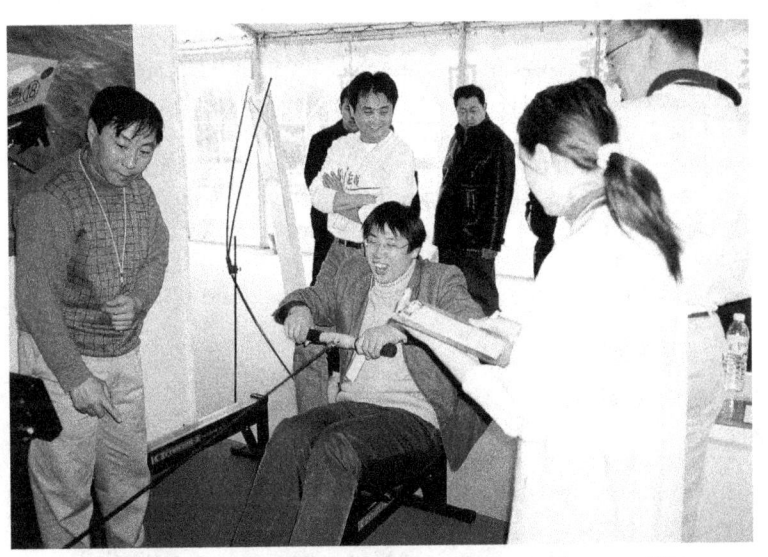

An excited participant in China's first amateur indoor rowing competition.

Sun Haibin was nominated for Sportsman of the Year 2002 in China and although he did not win, it was a great honour. A real 'from zero to hero' story and he deserves every bit of it. After graduating from Beijing Sports University he was employed by the university to help organise sports competitions. He now works for Zhang Jian. Sun Haibin and Cao Xinxin broke up shortly after the race and he was very upset, but got over it and now has a nice new girlfriend called Gong Wei.

In September 2002 Yu Zhen, who by then had left CCTV and was studying in England, and I went to Atlantic College where I gave a talk about the Yantu Project to the students and met Jin Zhou and Hui Wang. They both fit in well and are academically very talented, particularly in sciences. Jin won the prize for the best science and mathematics essay of the year and was invited to the Nobel Award ceremony in Stockholm in December 2002. He also qualified for the British Mathematics Olympiad and has already got offers from Harvard and Cambridge. Hui, having only just arrived, has qualified himself for the British Physics Olympiad and put on a play for the British Education Minister!

Jin later wrote to me:

> I now have friends from more than 60 countries all over the world. Meanwhile, I am also now one of the co-organisers of a China Project, organising about 30 Atlantic College students to teach English in a summer school in Sichuan Province in China for a month next summer vacation, to raise money for poor orphans to finish their education in China. For my whole life, I will always remember and appreciate everything in Atlantic College, a place to make a difference, both to myself and to the outside world.

And Hui wrote:

> I still remember the time when I received my enrolment with full scholarship letter from AC. It was so exhilarating that I hardly slept that night. I was just too excited to avoid all kinds of thoughts and fall asleep. What on earth does Atlantic College look like? I was still curious though as a matter of fact I had seen quite a few pictures of it, which were gorgeous. What will I experience there? How are other students from other countries? How can I communicate with them? Are they friendly enough to forgive my poor English and keep communicating with me? Quite interesting but meanwhile a little bit mysterious. I, as a teenager who had never been outside China, though had been taught a lot about the outside world, still it seemed thousands of light years away. Being abroad studying, which seems impossible to most of the Chinese now became practicable to me. Things are quite different in AC. It is a community as well as a formal college. Everything happens within the campus and no one stays disconnected to others as comprehensive housing system keeps everyone living and working together, regardless of different cultural backgrounds, which offers me a good chance to widen my vision by talking to other students. Besides academic subjects, extracurricular programmes also play a great part of my life. They give me satisfaction as well as chances of communicating. Yes, though it might not feel good to be thousands of miles away from home, still I consider it worth, as I have been learning what I have never learnt before and experiencing another sort of life, which will be unforgettable in my lifetime.

Postscript

It was great to meet Jin Zhou (left) and Hui Wang in person at Atlantic College in 2002, the two Mainland Chinese students we had raised scholarships for.

As a closing remark, should you have been inspired by our endeavours and would like to support United World Colleges then you can make donations to United World College through the following link: www.uwc.org/donate.

You can choose to give to one of the existing colleges or to the International Office. In any case, please write "Yantu" in the comments field, so that it is possible to track how many donations this second edition results in.

Alternatively (or better - in addition), buy a copy of this book for your employees, friends, and family. It would be great if book sale proceeds from this second edition exceed the USD1,500 from the first edition. Proceeds will be shared equally between the existing United World Colleges.

Finally, what is your dream?
Remember
if you really want to do it, then
anything is possible!

GO FOR IT!

Appendix

Yantu's Performance

In total we rowed 2,745 nautical miles or 5,084 kilometres. Our average speed was 2.02 knots or 3.74 kilometres per hour. The maximum distance we covered in one day was 69.4 nautical miles (128.5 km) and the minimum 6.0 (11.0 km) and our average was 48.5 (89.7 km).

The southern route we chose was 197 nautical miles (365 km) longer than if we had rowed straight across.

	Nm	Knots	Total Duration of Crossing		Km	Km/hour
Total	2,745.3	2.02		Total	5,084.2	3.74
Min	6.0	0.25	**56 days, 15 hours and 52 minutes**	Min	11.0	0.46
Max	69.4	3.10		Max	128.5	5.74
Avg	48.5	2.02		Avg	89.7	3.74

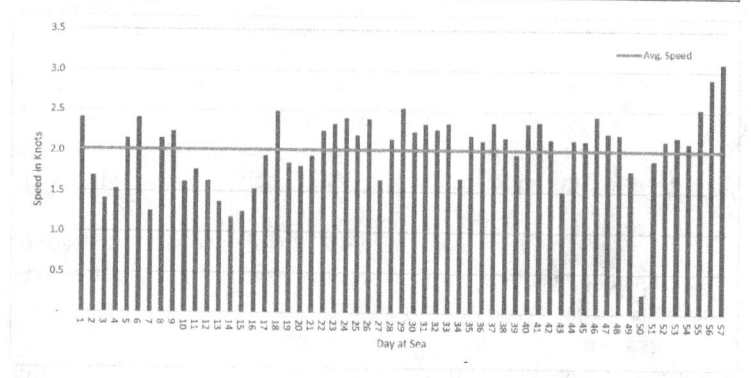

Appendix

Race Positions

Of the 36 boats at the start, 33 made it to the other side and only three withdrew. Of the finishing boats, 24 made it under the Race Rules and received a race placement. Nine boats were either resupplied with food, water, spare parts or received a tow and were unplaced.

Ward Evans Atlantic Rowing Challenge 2001 placings

#	Boat	Rowers	Duration
1	Telecom Challenge 1	Steve Westlake & Matt Goodman	42 days 2:16
2	Freedom	Patrick Weinrauch & Paul McCarthy	45 days 9:20
3	Win.Belgium	Bruno & Alain Lewuillon	49 days 4:40
4	Telecom Challenge 25	Steph Brown & Jude Ellis	50 days 7:00
5	Comship.com	Ian Anderson & Andy Chapple	50 days 15:11
6	Bright Spark	Tim Thurnham & Will Mason	50 days 16:22
7	Bruxelles Challenge	Pascal Hanssens & Serge van Cleve	55 days 13:32
8	**Yantu**	**Christian Havrehed & Sun Haibin**	**56 days 15:52**
9	Team Manpower	Richard White & Ian Roots	56 days 16:16
10	Atlantic Warrior	Rory Shannon & Alex Wilson	57 days 3:20
11	American Star	John Zeigler & Tom Mailhot	58 days 3:54
12	UniS Voyager	István Hajdu & Simon Walpole	58 days 9:50
13	EspritPME	Pierre Deroi & Jean Jacques Gauthier	59 days 1:11
14	Challenge Yourself	Scott Gilchrist & Peter Moore	60 days 14:14
15	Project Martha 2	Pedro Ripol & Francisco 'Pancho' Korff	61 days 7:29
16	Spirit of Jersey	Kerry Blandin & Paul Perchard	62 days 5:42
17	Keltec Challenger	Tim Humfrey & Jo Lumsdon	63 days 3:26
U	Onward	Dominic Marsh & Gary Fooks	63 days 5:52
18	McLlaid	Julian McHardy & Mark Williams	63 days 16:47
19	Team Nutri-Grain	Dominic & Crispin Comonte	64 days 5:13
U	La Gironde	Benjamin Marty & Oliver Villain	64 days 12:31
U	Linda	Alistair Smee & Chris Marett	69 days 13:9
20	Mrs D	Steve & Mick Dawson	70 days 10:12
21	Brunetto	Dugald Macdonald & David Mitchell	71 days 18:30
U	Human Rights	Denis Bribosia & Gregory Loret	73 days 12:45
22	The George Geary	Graham Walters & Michael Ryan	77 days 21:2
23	43<198> West	Damian West & Alex Hinton	79 days 21:41
U	Euskadi	Xabier Agote & Urko Mendiburu	83 days 14:58
24	This Way Up	Ian Chater & Tony Day	84 days 19:4
U	Uppsala.com	Niclas Mårdfelt & Rune Larsson	95 days 5:33
U	Domani 2	Rik Knoop & Michael Tuijn	102 days 16:15
U	Kaos	Ben Martell & Malcolm Atkinson	107 days 15:15
U	Troika Transatlantic	Debra (& Andrew) Veal	111 days 5:43
W	Dartmothian	David & Jason Hart	
W	Spirit of Worcestershire	Richard Wood & Rob Ringer	
W	Star Challenger	Jonathan Gornall & Dominic Biggs	
U		Unplaced	
W		Withdrawn	

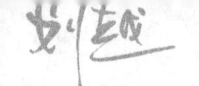

Yantu's Finances

Despite having written to many of China's most successful companies and a number or rich individuals we did not receive one cent of sponsorship from Mainland China towards scholarships or race costs. It all came from Hong Kong and international companies and individuals. I was disappointed about this, but Zhang Jian put it in perspective:

'All the bosses can still remember not having enough to eat, so they are not going to give money away. What you were trying to do was at least 10 years too early for Chinese companies.'

Zhang Jian has a point, but given the wealth some Mainland Chinese companies and individuals are enjoying, not sponsoring sounds to me a bit like a 'this is not how things are done in China' excuse. China is rapidly becoming international in its outlook and business and with that follows obligations of good corporate citizenship. So, when I launch my next Yantu Project, I will be back knocking on the doors of Chinese companies! I do not expect it to be easy to persuade Chinese companies to sponsor, but where there is a will there is a way. Anything is possible!

As can be seen from the table below, then proceeds from the sale of the first edition of this book resulted in an additional USD1,500 towards scholarships.

Appendix

Scholarship Account	USD
Scholarship donations by companies and individuals	88,620
Proceeds from 1. edition of book	1,500
Grand total	**90,120**
Cost per 2-year full scholarship	46,500
Total scholarships funded	**1.94**

Race Cost Account	
Race entrance fee to Challenge Business	16,623
Assembly of boat in China, including boat kit	14,633
Kitting out of boat in Hong Kong*	19,341
Pre-race travel for Sun Haibin	3,774
Safety & communications equipment	9,179
Food provisions	4,417
Freight to/from race	7,152
Rowers & family flight and accommodation	9,436
Marketing	2,576
Total race cost	**87,131**
Less	
Sponsored equipment**	6,239
Cash donations from companies and individuals	19,000
Sale of Yantu	20,313
Race cost assistance received	**45,552**
As % of total race costs	*52%*
Investment by Christian Havrehed	41,579

* Excludes manhours sponsored by Royal Hong Kong Yacht Club
** Mainly safety equipment sponsored by Viking and Stratos

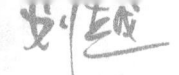

Yantu's Ownership

Yantu was built in 2000 by Luyang Boat Building Company, located in the countryside not far from the city of Shanwei in Guangdong Province, China. Luyang's facilities were too small to fit *Yantu* inside, so she was built outside under a garage roof.

She was the first ocean rowing boat to be built in Mainland China. Luyang was proud to build her and did so with great skill and dedication. As a result, *Yantu* is still going strong today.

She has by now completed five Trans-Atlantic rowing races as well as two Pacific ocean races, clocking up close to 20,000nm (35,000km) – equivalent to almost one time round the world, and 14 rowers have successfully put their lives in her hands, during their combined more than one year at sea (380 days).

Renamed *Row Aloha* she is currently holidaying in Hawaii where her current owner Todd is pondering rowing her from Hawaii to Australia in 2021.

Acquired	Boat name	Rowers	Nationality	Race	Distance	Start	Finish	Duration
2000	Yantu	Christian Havrehed / Sun Haibin	Denmark / China	2. Atlantic Rowing Race Tenerife - Barbados	~2700nm / ~5000km	7-Oct-01	3-Dec-01	56d 15h 52m
2002, Dec	Team Altitude	Christopher Hall / Richard Pullan	GB / GB	3. Atlantic Rowing Race La Gomera - Barbados	~2700nm / ~5000km	19-Oct-03	18-Dec-03	60d 13h 30m
2005	Mark 3 International	Robert Eustace / Peter Williams	GB / GB	5. Atlantic Rowing race La Gomera - Antigua	~2700nm / ~5000km	30-Nov-05	6-Feb-06	68d 01h 03m
2007	Ocean Summit	Neil Hunter / Scott McNaughton	GB / GB	6. Atlantic Rowing Race	~2700nm / ~5000km	2-Dec-07	7-Feb-08	67d 10h 10m
2009-2010	Ocean Summit	Stuart Burbridge / Rob Casserly	GB / GB	7. Atlantic Rowing race La Gomera - Antigua	~2700nm / ~5000km	4-Jan-10	27-Mar-10	82d
	Ocean Summit	??	GB	Did not use boat				
	Ocean Summit	Freddy Johnson	GB	Did not use boat				
2013, Jun	La Cignogne	Clément Héliot / Christophe Papillon	France / France	1. Great Pacific Race Monterey - Honolulu	~2500nm / ~4600km	9-Jun-14	23-Aug-14	75d 09h 25m
2014, Sep	Row Aloha	Todd Bliss / Rick Leach	US / US	2. Great Pacific Race Monterey - O'ahu, Hawaii	~2500nm / ~4600km	15-Jun-16	29-Jul-16	54d 22h 17m

Afterword – 19 years on

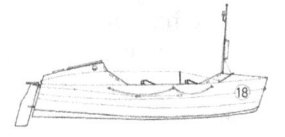

So what has become of Hui Wang and Jin Zhou, the two Mainland Chinese students we sponsored at Atlantic College? And what about Sun Haibin and myself?

Hui Wang / 王辉

Hui Wang graduated from Atlantic College in 2004 with maximum score in the IB and a full scholarship to study biochemical sciences at Harvard.

Unfortunately, on 7 October 2006, two years into Harvard, Hui was killed in a traffic accident at the age of 22. He had been hiking with friends in New York's Catskill Mountains when, on their way back to Harvard, their car swerved and suffered a full-frontal collision with an oncoming car. Three out of five in the car died. The remaining two recovered after months in hospital. The driver of the other car survived, too.

I have spoken at length to Hui's friends, including Jin, about how best to write about Hui in this Afterword, but there is no good way. When someone that talented, with everything going for him, dies in a senseless accident, it sucks. Hui was a single child, so not only did his parents lose their only child, they also lost the person who would look after them in old age. Hui came from a poor background and the change in living standard his education at Atlantic College and Harvard would have been able to bring his parents and immediate family would have been a dream come true for them. Instead everything turned into a nightmare.

Harvard enabled Hui's parents to come to America to collect his urn and Hui's friends collected money for the parents, but nothing can ever make up for the loss.

What is the lesson in this? That life is not fair? That we should always apply ourselves and be the best we can be because we never know when our time is up? If we could ask Hui, he would no doubt choose the latter.

Hui was not one to sit around brooding that life is not fair. Despite his quiet demeanour, he was a get up and goer, with a great appetite for life, sciences, photography, hiking, Japanese manga, and stimulating debates about anything. He had earnest curiosity and never complained. He was a humble and true scholar, as well as a good friend and listener, who was keen to make a difference by getting involved in the community.

We cannot bring Hui back, but his legacy inspires us. He, more than most, showed us that if you focus and follow your dreams, then anything is possible! Although Hui is no longer with us, he is not gone.

Afterword – 19 years on

Jin Zhou / 周进

In November 2019 I had the pleasure of catching up with Jin Zhou for the first time since 2002.

After Atlantic College Jin went to Harvard on full scholarship and graduated with a degree in Applied Mathematics. He then did what many Ivy League top graduates do; joined Wall Street where he traded stocks and bonds for a hedge fund. After four years he decided pursuing the Chinese dream was better than the American dream and he returned to China in 2011, where he now works in M&A helping a Chinese investment company buying up assets overseas.

When Jin got his scholarship to Atlantic College in 2001, the Chinese dream was to leave China to study and live overseas. 18 years later, Chinese still want to study overseas to gain international experience, but it is no longer a given that they want to remain abroad once they have completed their studies. Top Chinese talent can now find jobs within China, which pay at least as well as overseas jobs and with equally exciting, if not better, career prospects. Moreover, within China, the most attractive employers are no longer the foreign companies, it is the Chinese.

Think about that for a moment. A top Harvard graduate in mathematics decides to return to China instead of living the American dream because the Chinese dream is more appealing career-wise and pay-wise. What does that say about the outlook for the West? And for China?

Although China develops at a pace where you have to work hard simply not to fall behind, it is not all business for Jin. He has a passion for studying superstition in ancient China and is an avid scholar of fengshui and classic texts such as The Inner Canon of the Yellow Emperor and the Book of Changes. He has also found the time to get married and has two daughters.

Jin is an active member of the UWC alumni network in Hong Kong and Guangzhou. Through his work he has organised a CSR activity to interact with and sponsor autistic children in special needs schools in Hong Kong.

Afterword – 19 years on

Sun Haibin / 孙海滨

Sun Haibin with his wife Gong Wei and their two daughters.

Sun Haibin married his girlfriend Gong Wei and they now have two daughters. Him and I have kept in regular contact ever since our row. Growing up Sun Haibin did not have the means to buy bicycles, kayaks, wetsuits, lifejackets etc. to bring to sports events and that has always pained him. Because of this he has always dreamt about organising outdoor amateur adventure races where the organiser provides the necessary equipment and participants simply show up. In this way everyone can participate, not just those who can afford to buy the kit. Sun Haibin has managed to realise this dream!

In August 2013 Sun Haibin incorporated the company Kabrathon (www.kabrathon.com), which is short for Kayak – Bike – Run. The concept is similar to a Triathlon, except the swim, which many find hard, is exchanged for kayaking, and the distances are flexible depending on the venue. He has grown Kabrathon into a successful brand in China and he organises competitions throughout the country. As part of his outdoor-sports-for-all concept, he built an Exploration Camp by the Great Wall, complete with an artificial lake, where parents could take their children camping, rock climbing, kayaking etc. in a controlled environment. However, a few years ago, the camp was unfortunately expropriated, with little compensation. This was a great loss for Sun Haibin, but has since had more positive experiences with the Chinese system.

When Sun Haibin incorporated Kabrathon, he took care to register the Intellectual Property rights as well. This has proven useful, because although China has changed a lot, people will still attempt to copy your business or brand if you are successful. Kabrathon is no exception. One day Sun Haibin found out that some local operators had started organising Kabrathon-branded adventure races without his knowledge. He subsequently sued them – and won! He now receives a royalty fee each time a third party organises a Kabrathon-branded event. So that is an example of a completely different side of China, which we hear very little about in the West – a China where you can use the rule of law to enforce your IP rights.

... and they are off! The start of one of Sun Haibin's Kabrathon amateur competitions, which are becoming increasingly popular.

Afterword – 19 years on

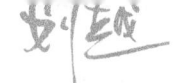

Christian Havrehed / 黄思远

Nikolaï, me, Véronique, and Victor. Victor is now at the Mahindra United World College in India.

As for myself, my family and I left China in 2013 and moved to Denmark. After 10 years in China my wife Véronique and I were ready for a break. Victor and Nikolaï were both born in Shanghai and it increasingly occurred to us that although the lifestyle was great and China is full of opportunities, China would never be home for them, so we decided to move to Europe to give them some roots and a better understanding of their French and Danish heritage. Nikolaï now studies at the French School in Copenhagen and Victor is at the United World College in India. Véronique and I have changed roles. She now works full time and I am a stay at home dad, adventurer, cross-cultural teambuilder, keynote and inspirational speaker (For how to book me see chapter 'Support Christian's adventures').

The adventure project I am currently working on involves retracing possible Chinese visits to America pre-Columbus. Imagine if the Chinese, like my Viking forefathers, visited America before Columbus! That would re-write the white man's History of Discovery. Completing this project will take years and will involve serious historical and archaeological research combined with various adventure trips and re-enactments of historical Chinese voyages, using ancient technology.

In 2021 it will be 20 years since Sun Haibin and I rowed across the Atlantic. We plan to mark this occasion by rowing from China to Japan in May/June 2021 in order to re-trace the voyage of Xu Fu, an alchemist who the First Emperor of China (the Qin Shihuang – whose tomb is at Xian with the Terracotta Army) sent to search for the Elixir of Immortality in the Eastern Seas in 210 B.C. Xu Fu and his fleet visited Korea and Japan, but no-one knows where he and the 3,000 youths he brought with him finally settled. Could they have made it to America? Respected sinologists, like the late Joseph Needham, think it could have been possible[2].

Xu Fu sailing into the Eastern Seas with 3,000 youths in 210 B.C..
Did they make it to America?

---

2   Joseph Needham. Science and Civilisation in China, Volume 4 - Physics and Physical Technology, Part 3 - Civil Engineering and Nautics, Cambridge University Press, 1971. Section 29 – Voyages and Discoveries - (viii) China and pre-Columbian America, page 553: *"Nevertheless, it may be almost equally likely that the story of Hsü Fu's [Xu Fu] disappearance conceals one voyage at least to the American continent".*

Afterword – 19 years on

There is a strong ocean current off the East coast of Japan, called the Kuroshio Current, which flows towards America. Washed out debris from the 2011 Tsunami in Japan has made it all the way across the Pacific Ocean onto the beaches in Oregon and California. Could Xu Fu and his crew have used the Kuroshio current to reach America?

Our rowboat will be technologically more advanced than what Xu Fu set out to sea in, but unlike Xu Fu's vessels we will not have any sails and will rely solely on rowing, so from that perspective we are at a disadvantage compared to Xu Fu. Although we will not be using a comparable vessel (at least not on this first voyage), we will still get a good understanding of the navigational challenges Xu Fu and his fleet will have encountered during their voyage – and if we are lucky might even find the Elixir of Immortality along the way. In the unlikely event we should fail to find the Elixir, then it will still be a great adventure where we will be bringing to life an interesting episode of China, Korea and Japan's shared history, whilst raising money for charity and awareness about United World Colleges.

The Chinese are rightly immensely proud of their long history, but few, Chinese and foreigners alike, know much about it, which is a great shame and I hope to change this with this new project.

In the 2004 Chinese edition of this book I explained in detail my plan to re-enact Xu Fu's voyage using an ancient vessel, i.e. like Thor Heyerdahl's *Kon-Tiki*, but so far no Chinese has caught on to the idea of bringing Chinese history to life through re-enactments. Like in 2001 when Sun Haibin and I became a catalyst for outdoor amateur sports, I hope this "Retracing possible Chinese visits to America pre-Columbus" project will become a catalyst for Chinese individuals to bring Chinese history to life in new and interesting ways for the benefit of Chinese and foreigners alike.

Who knows? Maybe 10 years from now two Chinese dressed in medieval clothes will show up on camels at the Vatican and when asked they will say: "We are re-enacting merchants Wang and Zhang's journey on the Silk Road in the 13th century. You know, the guys who brought Marco Polo back to China with them". This is obviously an imaginary example, but it illustrates a key point. Have you ever thought of the Silk Road involving Chinese coming to Europe or is your mental picture limited to Marco Polo going to China and coming back? In the West we tend to think of the History of Discovery as white men setting out to explore, but that is an ethnocentric view. Other countries explored, too. Those stories are just not well known in the West. I believe if these stories get told and accepted in the West, this will help promote tolerance and understanding between peoples, something still sorely needed in this day and age, in particular between China and the USA.

There are many positive elements encapsulated in "Retracing possible Chinese visits to America pre-Columbus" and the project has the potential to re-write the History of Discovery, but will it be able to attract Mainland Chinese funding? We'll see...

> You can read more about this adventure on
> www.yantu.com/adventures/retracing-chinese-visits-to-america-pre-columbus/
> and on Facebook
> @retracingchinesevisitstoamericaprecolumbus.

Support Christian's adventures

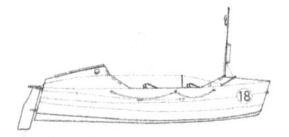

If you have enjoyed this book and like what Christian is doing, please support Christian undertake more adventures and write more books. This can be done in several ways;

- ✓ Make a donation towards his research & expedition costs;
- ✓ Hire Christian to Inspire at your company, association, Rotary Club etc;
- ✓ Enter into a corporate sponsorship agreement with Christian and visibly associate your company with his adventures, reaping returns through media exposure and goodwill;
- ✓ Promote Christian's adventures and services in your network.

Christian Havrehed has 20 years' experience in delivering unforgettable talks and team building, which challenge the status quo. He can deliver his services in English, Mandarin Chinese, German, and Danish. His books have so far been published in English, Chinese and Danish.

Prior to becoming a full-time adventurer and Chief Inspiration Officer, Christian worked 20 years in business. He spent four years in Business Development, eight years in Management Consulting, and eight years as a Managing Director. Past employers include L.E.K. Consulting in London as well as KPMG, Allianz, and VIKING Life-Saving Equipment in Hong Kong and Shanghai. He has renovated several houses and enjoys chopping firewood. In total, he has lived 20 years in China.

Christian is a United World College alumna. He holds a BA (Joint Hons.) in Chinese Language and Western Management Studies from Durham University, England, and an EMBA from Copenhagen Business School, Denmark.

Due to his solid business background, Christian is able to quickly grasp client issues and tailor make inspirational talks and / or team building to address these issues. He typically asks clients what three key messages they are trying to get across and then incorporates these sublimely into his talks or team building exercises. He also works strategically with clients to uncover what these focus areas should be to move the organisation ahead.

Christian uses his adventures to promote UWC values, cross-cultural understanding, and a different picture of China than what the Western mainstream media offers.

Christian is down to earth, approachable, and his clients rate him highly as an inspirational speaker and team builder.

> Client testimonies and how to book
> Christian for an event can be found at:
> https://yantu.com/speaking/
>
> Donations towards research &
> expedition costs (as well as designated
> charities) can be made here:
> https://yantu.com/donate/
>
> If you would like to discuss entering
> into a sponsorship agreement,
> please go to:
> https://yantu.com/contact/
>
> If you do not feel like going through
> the website then simply write to
> connect@yantu.com

Anything is Possible!